Modern Computational Finance

Modern Computational Finance

Scripting for Derivatives and xVA

ANTOINE SAVINE AND
JESPER ANDREASEN

WILEY

Published by John Wiley & Sons, Inc., Hoboken, New Jersey.
Published simultaneously in Canada.

For general information on our other products and services or for technical support, please contact our Customer Care Department within the United States at (800) 762-2974, outside the United States at (317) 572-3993 or fax (317) 572-4002.

Wiley also publishes its books in a variety of electronic formats. Some content that appears in print may not be available in electronic formats. For more information about Wiley products, visit our web site at www.wiley.com.

Library of Congress Cataloging-in-Publication Data

Names: Antoine Savine, author. | Jesper Andreasen, 1970– author.
Title: Modern computational finance : scripting for derivatives and XVA / Antoine Savine and Jesper Andreasen.
Description: Hoboken, New Jersey : John Wiley & Sons, Inc., [2022] | Includes bibliographical references.
Identifiers: LCCN 2021036994 (print) | LCCN 2021036995 (ebook) | ISBN 9781119540786 (hardback) | ISBN 9781119540816 (adobe pdf) | ISBN 9781119540793 (epub)
Subjects: LCSH: Finance—Mathematical models. | Finance—Computer simulation. | Finance—Data processing.
Classification: LCC HG106 .A556 2022 (print) | LCC HG106 (ebook) | DDC 332.01/5195—dc23
LC record available at https://lccn.loc.gov/2021036994
LC ebook record available at https://lccn.loc.gov/2021036995

Cover Design: Wiley
Cover Image: © kentarcajuan/Getty Images

SKY10029924_092121

Contents

My Life in Script by Jesper Andreasen

Antoine came to General Re Financial Products in London in 1998 with a lot of youthful spirit and many refreshing ideas. One of them was a financial payoff language called SynTech (Syntactic interpreter Technology). I am not sure that I was immediately convinced but when he connected it to a real model and priced some structures that we made up on the fly, I was hooked. I learned SynTech in hours but it took me months to figure out how it was put together, a process that forced me to learn structured programming in general and C++ in particular.

As always, it was a general struggle to keep up with the financial innovation, and constant re-coding of new payoffs was a painful and error prone process. We had been toying with a cocktail of Visual Basic for scripting of the payoffs and scenarios of future prices generated by C programs. However, the implementation was slow, model specific, hard to use, and generally more elephant than elegant.

SynTech, however, was easy to use with a readable simple Basic-like syntax and thoroughly built for speed and versatility. Scripts were pre-processed to cache and index static information for maximum pricing speed, and its API allowed a seamless hook-up with any dynamic model in our library. In fact, it took Antoine very little time to hook up SynTech to our then newly developed Libor Market Model.

Further, it was clear that SynTech could be extended to perform various life cycle tasks such as fixings and payments to limit the manual burden on the back office in the handling of exotic trades. SynTech implemented a clear separation between instrument and model.

From then on, I have only used scripting languages as the interfaces for all the models I have developed. And when I need to price something I rarely use anything else than naked scripting. The scripting languages that I have used and developed have all had a syntax very close to the original SynTech. The developments that I have done have mainly been underneath. Remarkably, SynTech's original 50 keywords still cover the ground.

SynTech was inspired by developments Antoine had seen at previous occupations, most notably the GRFN system at Paribas. Other young quants spread out from Paribas and seeded ideas and developments of scripting languages at other institutions, for example, at UBS, Commerzbank, Nikko-Salomon, Citi, and Reech Capital.

Paribas's GRFN developed by Guillaume Amblard and Societe Generale's OptIt by Jean-Marc Eber were as far as we know the first scripting languages that were used on an industrial scale. Jean-Marc Eber went on to set up the company Lexifi, which to this date supplies the industry with scripting technology.

GRFN and OptIt emerged in the mid-1990s. There were, for sure, other earlier attempts to develop scripting languages, most notably various uses of the LEX-YACC suite of tools for creating languages. Emmanuel Derman mentions such efforts in passing in *My Life as a Quant* and I have heard of similar experiments at JP Morgan. However, to my knowledge, none of these made it to large-scale production.

In 2001, I went to Bank of America in London, where I ganged up with James Pfeffer and developed Thor. The main innovation of Thor was the use of the visitor pattern technology that later was instrumental in using scripting as the backbone for xVA and regulatory calculations. The main idea of the visitor pattern is to have visitors that visit the cash flows for different purposes: pre-processing, valuation, dependency graph generation, compression, decoration, and so on. The diverse use of visitors shows that development of your own scripting language is necessary. Python or C# can not be visited.

In 2002–2005 I worked for Nordea in Copenhagen where we developed a scripting language called Loke. The main innovation during this period was the integration of American Monte-Carlo techniques with Loke, including upper-bound techniques implemented by Ove Scavenius.

In 2005–2008 I was back with Bank of America, where I found the Thor language to be heavily used by the structured interest rate and equity desks. Often they used a macro language called Sif, developed by Mohan Namasivayam, to generate Thor scripts.

At Danske Bank, which I joined in 2008 and left in 2018, the scripting language is called Jive. Jive is used for everything traded at Danske Bank: from vanilla swaps to mortgage bonds to equity exotics to regulatory capital. Jive is even used in the yield curve calibration. This is the only bank that I know of that uses a scripting language so consistently. Developments have mainly concentrated on aggregation, compression, and decoration used for xVA and regulatory calculations. Here, we have heavily leveraged on the visitor pattern technology. But we have also done significant work on Automatic Adjoint Differentiation (AAD), multi-threading, branching simulations, and fuzzy logic to stabilize risk calculations. Antoine joined Danske in 2013, and since then he has played a key role in these developments. In 2015, we received *Risk Magazine*'s In-House Risk System of the Year award for our xVA system. The AAD and the speed of the calculations achieved by the xVA system have received a lot of attention in the quant community, but the scripting and visitor pattern technologies are actually the unsung heroes that have made all of this possible.

To document how we did xVA on an iPad Mini without thoroughly describing our approach to scripting would be wrong and not give the reader the full picture. I was therefore very happy when Antoine told me he was going to write a book on scripting. This would give us the opportunity to finally get our work on scripting languages documented and pave the way for fully documenting our xVA system.

There are a number of reasons why the story of scripting has not been told before. Among these:

- It's not mathematics, but software design, which is actually something that we do every day but not something that we usually write about.
- It's a complex and relatively long and winding story that cannot easily be summoned in a few punchlines. On top of this it contains subjects that most people haven't even heard about, such as visitors and pre-processors.
- It's a hard sell to change conventional wisdom that scripting is only for exotics and has no relevance in post-crisis finance. Actually, scripting is more relevant now than ever, because banks are under tough regulatory and cost pressures.
- It's very C++ and as such not following the current trend of lighter languages such as Matlab and Python backed by GPUs running parallel instructions in C.
- The subject is still very much alive: whenever we start documenting our efforts, we always get new ideas that we can do with scripting, which in turn tends to take away focus from documenting past glories.

My next employer is Saxo Bank and the scripting language there will be Jife.

Jesper Andreasen, June 2018

PART I

A Scripting Library in C++

Introduction

This part leads readers through the development steps of the scripting library provided in our source repository.

A transaction consists of a number of events occurring on given dates in the future. This is where a payment or coupon is fixed, a discrete barrier is monitored, or an exercise takes place. This is also where a path-dependency is updated with reference to the simulated variables on that future date.[1] Future dates when such events take place are called *event dates*.

As an illustration, we consider a simplified version of the popular *autocallable* transaction. It pays a high coupon (say 10%) if some index (say S&P500) increased during the first year of its life. In this case, it pays the notional redemption together with the coupon of 10% and terminates at the end of year 1. Otherwise, it may still deliver a 20% coupon (10% per annum) at the end of year 2 provided the index overall increased over the two years. In this case, it repays the notional together with the 20% coupon and terminates at the end of year 2. If not, it is given a last chance on year 3, provided the underlying index increased over the three years. If this is the case, the investor receives the redemption + 30% at the end of year 3. If not, only 50% of the notional is repaid. It is easy to see that the investor implicitly sells a (somewhat exotic) option on what may appear as a low probability event (index decreasing over one, two, and three years) in exchange for a high coupon in a low-yield environment, which explains the success of this structure.[2]

---

[1]Continuous barriers are somehow outside of this logic and require specific support that is not discussed. Support for early exercises is discussed in chapter 15.

[2]This is an overly simplified version of the autocallable product and one that is purposely structured to illustrate the implementation and application of scripting. It is significantly different from actual autocallable transactions, which can also be scripted with our language, but in a more complicated manner.

This product may be scripted as follows (say today is 1 June 2020)[3]:

01Jun2020	vRef=spot() vAlive=1

01Jun2021	if spot() > vRef then
	prd=110
	vAlive=0
	endIf

01Jun2022	if spot() > vRef then
	if vAlive=1 then prd=120 endIf
	vAlive=0
	endIf

01Jun2023	if vAlive=1 then
	if spot() > vRef then prd=130 else prd=50 endIf
	endIf

We have four events on four event dates:

1. Today, we set the reference to the current spot level and initialize the *alive* status to 1.
2. Year 1, we check whether the spot increased, in which case we pay redemption + 10% and die.
3. Year 2, we check that the spot overall increased over two years. In this case, provided we survived year 1, we pay redemption + 20% and die.
4. Year 3, provided we survived the first two, we check if the spot overall increased. In this case we pay redemption + 30%. If not, we repay 50% of the notional.

We see that our language must at the very minimum support numbers, arithmetic operations, and conditional statements. We know we also need some mathematical functions like *log* or *sqrt* and some financial functions such as a multi-argument *min* and *max*. Critically, we must be able to read, write, and compare variables and access the simulated market with a *spot()* keyword that references the fixing of the underlying asset on the corresponding event date. This is a simple language, similar to Python, that supports

[3]We make some arbitrary choices regarding the syntax of our scripting language. Readers can easily implement their own preferred grammar once they are comfortable with the idioms explained here.

only the constructs necessary for the description of financial cash-flows, for which it provides some specific keywords.

The language considers as a variable anything that starts with a letter and is not otherwise a keyword. We used the variables *vRef*, *vAlive*, and *prd* in our example. Evidently, ancillary variable names don't *have* to start with the letter *V*; this is only for clarity.

Products are variables. The language makes no difference between products and ancillary variables; only users do. Our example actually scripts three products: *prd* obviously; *vRef*, which pays the spot fixing today and is worth today's underlying asset price; and *vAlive*, which is worth 1 at maturity if the product survived the first two years, 0 otherwise. Its value is the (risk-neutral) probability of surviving to year 3. All variables may be seen as products, although, in general, the values of ancillary variables are disregarded in the end. In chapter 5, we will implement other means of distinguishing products from helper variables, and actual payments from assignments, with the keyword *pays*.

To value a script in the context of path-wise Monte-Carlo simulations means to evaluate it against a number of scenarios, each scenario providing different values for *spot* on the event dates in 1y, 2y, and 3y. For every such scenario generated by the model, we evaluate the script and record the final value of all its variables. In the end, we average those final variables' values across scenarios to estimate the values of the corresponding products. If we also require risk sensitivities, we compute the derivatives of the final variable values to changes in the model parameters. Evidently, the derivatives of prices are averages of the path-wise derivatives, which permits an efficient path-wise computations of sensitivities, in particular with adjoint propagation. See for instance Giles and Glasseman's "Smoking Adjoints," which introduced adjoint techniques to finance [14], and our publication [27], which explains automatic adjoint differentiation (AAD) and provides professional differentiation code.

We remind the reader, however, that evaluation (including of sensitivities) is only one thing we can do with the script. The purpose of this library is to parse scripts into visitable data structures and implement a framework that enables all types of visitors, not only the evaluator, to traverse and manipulate scripts in many ways.

We split the implementation in five steps.

First, we describe in chapter 2 the data structure for the internal representation of the script, ready for evaluation and other forms of visits. We will use *expression trees* as a data structure, and we describe these in detail. The discussion and code for the actual *parsing* (that turns a text script into a collection of expression trees) is left to the appendix.

Then, we introduce in chapter 3 the *evaluator* that values expression trees during simulations, *pre-processors* that optimize evaluation before simulations, and other *visitors*, objects that traverse expression trees and perform calculations and actions depending on the visited node, while maintaining internal state. We explain the *visitor pattern*, a common framework for all types of visitors, which encapsulates traversal logic and makes the development of specific visitors particularly simple.

Third, in chapter 4, we bring the pieces together and develop a (very) basic model to test the scripting library.

Fourth, we improve our framework with the addition of the keyword *pays* in chapter 5 and take this opportunity to illustrate how a core extension is made to the language.

CHAPTER 1

Opening Remarks

INTRODUCTION

In the early stages of derivative markets, dedicated models were typically put together to value and risk manage new transaction types as they emerged. After Black and Scholes [5] published in 1973 a closed-form formula for European options under a constant volatility assumption, alternative models—like the Cox-Ross-Rubinstein binomial tree [7] in 1979, later replaced by more efficient finite difference grids—were developed to value American options under the same assumptions.

As trading in derivatives matured, the range of complex transactions expanded and models increased in complexity so that numerical methods became necessary for all but the simplest vanilla products. Models were typically implemented in terms of finite difference grids for transactions with early exercise and Monte-Carlo simulations for products with path-dependency. Notably, models increased in dimension as they grew in complexity, making grids impractical in most cases, and Monte-Carlo simulations became the norm, with early exercises typically supported by a version of the Longstaff-Schwartz regression-based algorithm [22]. Sophisticated models also had to be calibrated before they were used to value and risk manage exotics: their parameters were set to match the market prices of less complex, more liquid derivatives, typically European calls and puts.

Most of the steps involved—calibration, Monte-Carlo path generation, backward induction through finite difference grids—were independent of the transactions being valued; therefore, it became best practice to implement models in terms of generic numerical algorithms, independently of products. Practitioners developed modular libraries, like the simulation library of our publication [27], where transactions were represented in separate code that interacted with models to produce values and risk sensitivities.

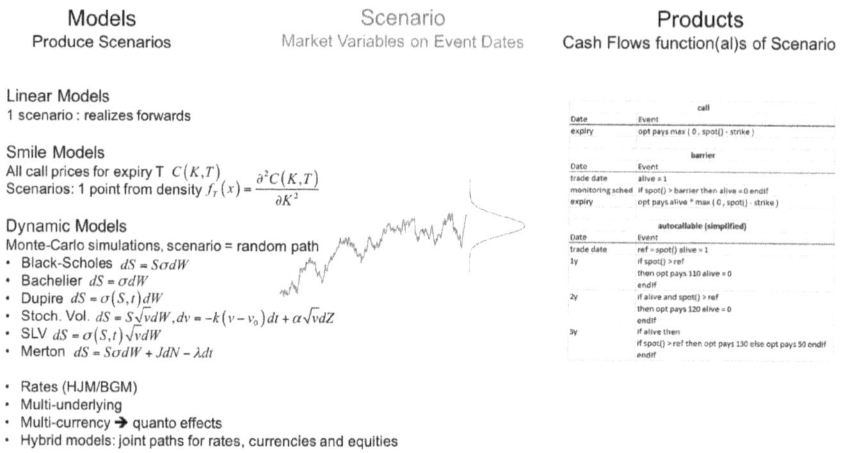

Models — Produce Scenarios

Scenario — Market Variables on Event Dates

Products — Cash Flows function(al)s of Scenario

Linear Models
1 scenario : realizes forwards

Smile Models
All call prices for expiry T $C(K,T)$
Scenarios: 1 point from density $f_T(x) = \frac{\partial^2 C(K,T)}{\partial K^2}$

Dynamic Models
Monte-Carlo simulations, scenario = random path

- Black-Scholes $dS = S\sigma dW$
- Bachelier $dS = \sigma dW$
- Dupire $dS = \sigma(S,t)dW$
- Stoch. Vol. $dS = S\sqrt{v}dW, dv = -k(v - v_0)dt + \alpha\sqrt{v}dZ$
- SLV $dS = \sigma(S,t)\sqrt{v}dW$
- Merton $dS = S\sigma dW + JdN - \lambda dt$

- Rates (HJM/BGM)
- Multi-underlying
- Multi-currency → quanto effects
- Hybrid models: joint paths for rates, currencies and equities

call

Date	Event
expiry	opt pays max (0 , spot() - strike)

barrier

Date	Event
trade date	alive = 1
monitoring sched	if spot() > barrier then alive = 0 endif
expiry	opt pays alive * max (0 , spot() - strike)

autocallable (simplified)

Date	Event
trade date	ref = spot() alive = 1
1y	if spot() > ref then opt pays 110 alive = 0 endif
2y	if alive and spot() > ref then opt pays 120 alive = 0 endif

3y	if alive then
	if spot() > ref then opt pays 130 else opt pays 50 endif
	endif

However, at that stage, dedicated code was still written for different families of transactions, and it was necessary in order to add a new product to the library, to hard code its payoff by hand, compile, test, debug, and release an updated software.

The modular logic could be pushed one step further with the introduction of *scripting languages*, where users create products dynamically at run time. The user describes the schedule of cash-flows for a transaction with a dedicated language specifically designed for that purpose, for example:

STRIKE	100
01Jun2021	opt pays max(0, spot() - STRIKE)

for a 1y European call with strike 100, or

STRIKE BARRIER	100 120
01Jun2020	vAlive = 1
Start: 01Jun2020 End: 01Jun2021 Freq: weekly	if spot() > BARRIER then vAlive = 0 endIf
01Jun2021	opt pays vAlive * max(0, spot() - STRIKE)

for the same call with a 120 (weekly monitored) knock-out barrier.[1]

The scripts are *parsed* into *expression trees*, and *visited* by an *evaluator*, a particular breed of *visitor*, who traverses the trees, while maintaining the internal state, to compute payoffs over the scenarios generated by a model:

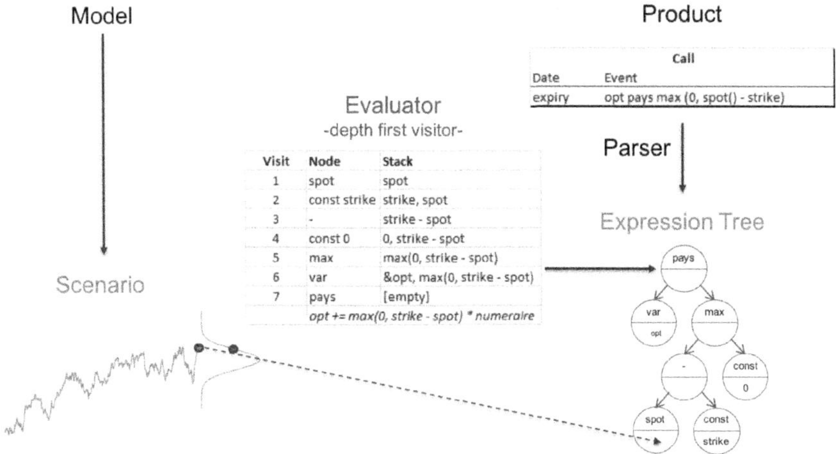

All of this is explained in deep detail with words and code in part I.

With scripting, finance professionals were able to create and modify a product on the fly, while calculating its price and risk sensitivities in real time. The obvious benefits of such technology quickly made it a best practice among key derivatives players and greatly contributed in itself to the development of structured derivatives markets.

---

[1]Depending on scripting grammar, that barrier could also be written *in functional form*:

STRIKE BARRIER	100 120
01Jun2020	append (vFixings, spot())
Start: 01Jun2020 End: 01Jun2021 Freq: weekly	append (vFixings, spot())
01Jun2021	opt pays (max(vFixings) < BARRIER) * max(0, spot() - STRIKE)

We develop a classic, procedural scripting language in what follows, although readers can easily apply the logic and code of this book to design a *functional* scripting language like in the script above, or implement the grammar of their choice.

Early implementations, however, suffered from an excessive performance overhead and a somewhat obscure syntax that made scripting inaccessible to anyone but experienced quantitative analysts and traders. Later implementations fixed those flaws. The modern implementation in this publication comes with a natural syntax, is accessible to non-programmers,[2] and its performance approaches hard-coded payoffs.

This publication builds on the authors' experience to produce a scripting library with maximum scope, modularity, transparency, stability, scalability, and performance.

Importantly, our implementation transcends the context of valuation and sensitivities; it offers a consistent, *visitable* representation of cash-flows that lends itself to a scalable production of risk, back-testing, capital assessment, value adjustments, or even middle office processing for portfolios of heterogeneous financial transactions. We also focus on performance and introduce the key notion of *pre-processing*, whereby a script is automatically analyzed, prior to its valuation or risk, to optimize the upcoming calculations. Our framework provides a representation of the cash-flows and a way of working with them that facilitates not only valuation but also pre-processing and any kind of query or transformation that we may want to conduct on the cash-flows of a set of transactions.

Scripting makes a significant difference in the context of xVA, as explained in part V, and more generally, all regulatory calculations that deal with multiple derivatives transactions of various sophistication, written on many underlying assets belonging to different asset classes. Before xVA may be computed over a netting set,[3] all the transactions in the netting set must be aggregated. This raises a very practical challenge and a conundrum when the different transactions are booked in different systems and represented under different forms. Scripting offers a consistent representation of all the transactions, down to their cash-flows. Scripted transactions are therefore naturally aggregated or *manipulated in any way*. A key benefit of scripted cash-flows is that scripts are not black boxes. Our software (more precisely, the visitors implemented in the software) can "see" and analyze scripts, in order to aggregate, compress, or decorate transactions as explained in part V, extract information such as path-dependence or non-linearity and select the model accordingly, implement automatic risk smoothing (part IV),

---

[2]We develop a scripting language similar to a high-level language like Python, not a programming language like C++. Classic constructs like variables, loops, and conditions are part of the language, as well as special keywords for the description of cash-flows, such as spot(), which references simulated market prices. Of course, this is one particular choice and different rules are implemented with minor modifications, for example, a C/C++ inspired syntax or a functional grammar without conditionals or variables.

[3]All transactions against a counterparty.

or analyze a valuation problem to optimize its calculation. Our library is designed to facilitate all these manipulations, as well as *those we haven't thought about yet.*

The purpose of this publication is to provide a complete reference for the implementation of scripting and its application in derivatives systems to its full potential. The defining foundations of a well-designed scripting library are described and illustrated with C++ code, available online on:

https://github.com/asavine/Scripting/tree/Book-V1

Readers will find significant differences between the repository code and the code printed in the book. The repository has been undergoing substantial modernization and performance improvements not covered in this edition of the text. *Make sure you connect to the branch Book-V1, not the master.* Besides, the code base evolves throughout the book and the online repository contains the final version. It is advisable to type by hand the code printed in the text rather than rely on the online repository while reading the book.

This code constitutes a self-contained, professional implementation of scripting in C++. It is written in standard, modern C++ without any external dependency. It was tested to compile on Visual Studio 2017. The library is fully portable across financial libraries and platforms and includes an API, described in section 3.7, to communicate with any model.

The code as it stands can deal with a model of any complexity as long as it is a single underlying model. It works with the Black and Scholes model of [5] and all kind of extensions, including with local and/or stochastic volatility, like Dupire's [9] and [10], or single underlying stochastic volatility models.[4] The library cannot deal with multiple underlying assets, stochastic interest rates, or advanced features such as the Longstaff-Schwartz algorithm of [22]. It doesn't cover the management of transactions throughout their life cycle or deal with past fixings. All those features, necessary in a production environment, would distract us from correctly establishing the defining bases. These extensions are discussed in detail in parts II and III, although not in code.

Our online repository also provides an implementation of Fuzzy Logic for automatic risk smoothing, an excel interface to the library, a tutorial for exporting C++ code to Excel, a prebuilt xll, and a demonstration spreadsheet.

The C++ implementation is discussed in part I, where we explore in detail the key concepts of expression trees and visitors. We show how they are implemented in modern C++ and define the foundations for an efficient,

[4]Stochastic volatility models are covered in many textbooks, including [21], [13], [3], or our lecture notes [26].

scalable scripting library. We also provide some (very) basic self-contained models to test the library, although the library is model agnostic by design and meant to work with any model that implements an API that we specify. For clarity, the code and comments related to parsing (the transformation of a text script into an expression tree) are moved from the body of the text into an appendix.

We discuss in part II the implementation of some basic extensions, and in part III more advanced features like interest rates and multiple currencies and assets. We discuss the key notion of indexing simulated data. Indexing is a special flavor of pre-processing, crucial for performance. We also discuss the support for LSM, the regression-based algorithm designed by Carriere and Longstaff-Schwartz in [6] and [22] to deal with early exercise in the context of Monte-Carlo simulations, and later reused in the industry in the context of xVA and other regulatory calculations. These parts include extensive guidance for the development of the extensions, but not source code.

The rest of the publication describes some applications of scripting outside the strict domain of pricing and demonstrates that our framework, based on visitors, can resolve many other problems.

Part IV shows how our framework can accommodate a *systematic* smoothing of discontinuities to resolve the problem of unstable risk sensitivities for products like digitals or barriers with Monte-Carlo simulations. Smoothing consists of the approximation of the discontinuous payoffs by close continuous ones, like the approximation of digitals by tight call spreads. Bergomi discusses and optimizes smoothing in [4]. Our purpose is different. We demonstrate that smoothing can be abstracted as a particular application of *fuzzy logic*. This realization leads to an algorithm to systematically smooth not only digitals and barriers but also any payoff, automatically. The practical implementation of the algorithm is facilitated by the design of our scripting library. For clarity, the source code is not provided in the body of the text, but it is provided in our online repository.

Part V introduces the application to xVA, which is further covered in our upcoming dedicated publication. The code for xVA calculations is not provided.

1.1 SCRIPTING IS NOT ONLY FOR EXOTICS

While scripted payoffs are now part of the standard toolkit for the structuring of derivatives, it does not appear to be widely used yet in the wider context of xVA, CCR, VAR, FRTB, capital charges, or other related calculations. This is an unfortunate oversight. Market risk, value adjustments, and capital calculations over large heterogeneous portfolios of transactions are best conducted when all cash-flows from all transactions are represented in a consistent manner, which the software can understand, analyze, and otherwise manipulate. This is exactly what scripting offers, for a small performance cost if implemented correctly.

Further, a CVA (similarly to other xVAs and other regulatory calculations) is a real option that a bank gives away any time it trades with a defaultable counterparty. That option is a put on the netting set, contingent to default. It is therefore an exotic option with a complex payoff written on the sum of all the cash-flows of all the transactions in a netting set. This is however still an option, and as such, it can be scripted like any other transaction. Scripting is ideally suited in this context, both for the representation of the cash-flow of the underlying transactions *and* the description of the value adjustments themselves. We explore the details in part V and will extend that discussion in our upcoming publication dedicated to xVA.

As exotic transactions fell out of favor in the aftermaths of the 2008 global crisis, some practitioners believe that scripting lost relevance. That could not be more wrong. From the reasons enumerated, scripting is more relevant now than ever, not only for exotics but more importantly today as the centerpiece of correctly designed systems for the calculation of risks, value adjustments, and capital charges for large heterogeneous portfolios.

We therefore believe that scripting is not limited to an *external* interface for users to conveniently create transactions on the fly. Scripting does offer such convenience, but its main purpose is to represent transactions *internally* as collections of cash-flows consistently accessible to all components in a system. Code (actually, visitors that are part of the code) can "see" the cash-flows and manipulate them in a number of ways. Evaluation is just one such manipulation. The design of our scripting library in part I is driven by such considerations.

1.2 SCRIPTING IS FOR CASH-FLOWS NOT PAYOFFS

What did lose relevance is scripting for exotics only. Scripting must be rethought today, no longer as a means of valuing complex transactions but as a way to represent cash-flows so that these cash-flows may be *visited* from various components of system and analyzed and modified in various ways, including, but not limited to, valuation. For this purpose, it is essential that it is the actual cash-flows that are scripted and not option payoffs.

For example, a receiver swaption with physical settlement would frequently be represented as a European option on the PV of the underlying swap at the exercise date. This would produce a script resembling:

	swaption pays max(0,
exerciseDate	swapPv (startDate, endDate, fixCpn, fixFreq, fixBasis, floatRef))

Such shortcut may be relevant for valuation,[5] because it is sometimes economically equivalent to receive the PV of the cash-flows on the exercise date, or become entitled to these cash-flows, paid later on their respective payment dates. But this shortcut only works in particular contexts and with particular models. When credit risk is considered, it is no longer equivalent to exercise into the swap or to cash settle the option. In particular, this script is not correct in the context of CVA.

The correct script for a swaption must describe the schedule of the cash-flows of the underlying swap and use a *proxy* to estimate the PV of the future cash-flows of the swap at the exercise date to make an exercise decision. A proxy is an estimate of the transaction's value on the exercise date as a function of the model's state variables on that date. Proxies are typically produced by regression, in a process designed by Carriere [6] and Longstaff-Schwartz [22], and discussed in chapter 15 and, more extensively, in our upcoming, dedicated publication.

Start: start DateEnd: end DateFreq: float FreqFixing: start-2bd	if vAlive = 1 then swaption pays -libor(StartPeriod, EndPeriod, floatBasis, floatRef) * cvg(StartPeriod, EndPeriod, floatBasis) on EndPeriod endIf
Start: startDate End: endDate Freq: fixFreq Fixing: start-2bd	if vAlive = 1 then swaption pays fixCpn * cvg(StartPeriod, EndPeriod, fixBasis) on EndPeriod
tradeDate	vAlive = 0
exerciseDate	if PV(swaption) > 0 then vAlive = 1 endIf

When we script payoffs, we lose the information of how they are pieced together out of the actual cash-flows. That may seem irrelevant for valuation, so it may be tempting to script payoffs directly, and that may even be thought of as a performance improvement, but, in a wider context where it is necessary to read the actual cash-flows and, for example, figure out how they are fixed and when they are paid, then such information is no longer available. Therefore, it is best practice, in modern finance, to always script the raw cash-flows and leave the optimization to the implementation.

A well-designed scripting implementation should provide facilities to extract information out of the scripted transactions, such as the dates

[5]It is indeed relevant for *calibration*.

when payments may occur, the collection of contingent and non-contingent (deterministic) payments, the identification of European, path-dependent and/or early exerciseable cash-flows, the dependency of cash-flows to underlying assets and market variables, the dependencies between different components of a transaction or portfolio, and much more. Such information may help optimize the setup of the model before it runs valuation. It can also be used for middle office processing, back-testing, or forward-testing like for FRTB or PRIIPS regulatory calculations. Smoothing with fuzzy logic (part IV) requires access to raw cash-flows and may fail if additional (sharp) logic is included in the script. Visitors may even modify the script after parsing, for example, to compress a large amount of heterogeneous cash-flows into a "super-swap" that pays the combined cash-flows from multiple transactions on evenly spaced dates, or decorate a set of cash-flows with the "payoff" of an xVA (part V), but only if the actual, raw cash-flows are scripted without additional shortcuts.

The key notion here is that a script is not a black box to be valued against scenarios, but a data structure that holds cash-flows and is open to *visitors*. Valuation is only one form of visit among a multitude of others that extract information from cash-flows and modify them in various ways. And for that purpose, it is necessary to script the raw cash-flows only.

1.3 SIMULATION MODELS

In a strict valuation context, the computation of the value and risk sensitivities of a script requires a model. Models are the primary focus of mathematical finance and are discussed in a vast number of textbooks, publications, and lectures. The clearest and most comprehensive description of the design and implementation of derivative models can be found in the three volumes of [20].

This publication is not about models. We develop a *model agnostic* scripting library that is written independently of models and designed to communicate with all models that satisfy an API discussed in section 3.7 (where we show that the API can also be *added* to some existing models with an adapter class). For our purpose, a model is anything that generates *scenarios*, also called *paths* in the context of Monte-Carlo simulations, that is, the value of the relevant simulated variables (essentially, underlying asset prices) on the relevant event dates (when something meaningful is meant to happen for the transaction, like a fixing, a payment, an exercise, or the update of a path dependency). Our library consumes scenarios to evaluate payoffs, without concern of how those scenarios were produced.

Scripting libraries are also typically written to be usable with a number of numerical implementations, either by forward induction like

Monte-Carlo simulations or backward induction like finite difference grids. In order to avoid unnecessary confusion, our book focuses on forward induction with Monte-Carlo simulations, by far the most frequently used valuation context today. To further simplify our approach, we only consider *path-wise* simulations. This means that simulations are provided one path (for all event dates) at a time. The model is, to us, an abstract object that generates multiple scenarios (possible evolutions of the world) and communicates them sequentially for evaluation.

An alternative that we don't consider is step-wise simulation, where all paths are computed simultaneously, date by date. The model first generates all the possible states of the world for the first event date, then moves on to the second event date, and so on until it reaches the final maturity. Step-wise simulation is natural with particular random generators, control variates (when paths are modified after simulation so that the expectation of some function matches some target), and, more generally, calibration inside simulations as in the *particle method* of [16].

The concepts and code in this book apply to path-wise Monte-Carlo simulations, but they can be extended to support step-wise Monte-Carlo and backward induction algorithms.[6] This is however not discussed further. Note that path-wise Monte-Carlo includes deterministic models for linear products (these models generate a single path where all forwards are realized) and numerical integration for European options (where each path is a point in the integration scheme against the risk neutral probability distribution of the relevant market variable).

Monte-Carlo simulations are explained in detail in [19], [15], and the dedicated chapters of [20], as well as our publication [27], which provides a framework and code in C++, and focuses on a parallel implementation.

We refer to the figure on page 15 for an illustration of the valuation process. The model produces a number of scenarios; every scenario is consumed by the evaluator to compute the payoff over that scenario. This is repeated over a number of paths, followed by an aggregation of the results. The evaluator is the object that computes payoffs (sums of cash flows normalized by the numeraire, which is also part of the scenarios, as described in chapter 5) as a function of the simulated scenarios. The evaluator conducts its work by traversal of the expression tree that represents the script, while maintaining an internal stack for the storage of intermediate results. We describe the evaluator in detail in section 3.6.

The evaluation of the script over a path is typically executed a large number of times; therefore, the overall performance is largely dependent on its optimization. This is achieved by many forms of *pre-processing*.

[6]We can accommodate step-wise MC in a trivial, if memory-inefficient manner: store the paths generated step-wise and deliver them path-wise to the script evaluator.

1.4 PRE-PROCESSING

Pre-processing is the key optimization that makes the valuation of scripts (almost) as fast as hard-coded payoffs. Where performance matters most is in the code repeated for a large number of simulations. Monte-Carlo developers know that maximum performance is achieved by moving as much work as possible before simulations take place: pre-allocation of working memory, transformation of data, pre-calculation of quantities that don't directly depend on the simulated variables. We want to perform as much work as possible once, before simulations start, so that the subsequent work conducted repeatedly over scenarios is as limited and as efficient as possible.

Pre-processing is not only about pre-computing parts of subsequent evaluations. It is not so much about the optimization of the *arithmetic* calculations performed during simulations. It is mainly about moving most of the *administrative*, logistic overhead to processing time. CPUs perform mathematical calculations incredibly fast, but the access of data in memory may be slow if not carefully dealt with. With hard-coded payoffs (payoffs coded in C++), it is the compiler that optimizes the code, putting all the right data in the right places in the interest of performance. With scripting, the payoff, as a function of the scenario, is built at run time. The compiler cannot optimize this function for us. This is our responsibility. This is where the pre-processors come into play.

A special breed of pre-processors called *indexers* arrange data for the fastest possible access during simulations, with a massive performance impact. For instance, when some cash-flow is related to a given Libor rate of a given maturity, the indexer "informs" the evaluator, before simulations start, *where* in pre-allocated memory that Libor will live during simulations, so that, at that point, the evaluator reads the simulated Libor there, directly, without any kind of expensive lookup.

Most noticeable benefits are achieved in the context of interest rates (and multiple assets) discussed in section III. The dates when coupons are fixed or paid and the maturities of the simulated data such as discount factors and Libors are independent of scenarios. The *value* of these simulated variables may be different in every scenario, but their *specification*, including maturity, is fixed. Pre-processors identify what rates are required for what event dates, and perform memory allocation, indexing, and other logistics before simulations start. During simulations, where performance matters most, the model communicates the values of the simulated data in a pre-indexed array for a fast, random access.

This is covered and clarified in part III, but we introduce an example straight away. Consider a script for a caplet:

```
            caplet pays 0.25 *
01Jun2021   max( 0, libor( 03Jun2021, 3m, act/360, L3) - STRIKE)
            on 03Sep2021
```

The payoff of this caplet on 03Jun2021 is:

$$0.25max(0, libor(03Sep2021) - STRIKE)\,discount(03Sep2021)/numeraire$$

The responsibility of the model is to generate a number of joint scenarios for the Libor and discount.[7] The responsibility of the evaluator is to compute the payoff above from these two values.

While this is all mathematically very clear, a practical implementation is more challenging. The product is scripted at run time, so we don't know, at compile time, what simulation data exactly the evaluator expects from the model. Dedicated data structures must be designed for that purpose, something like (in pseudo-code):

```
Scenario := vector of SimulData per event date
SimulData :=  numeraire, vector of libors, vector of discounts, ...
```

In our simple example, we have one event date: 01Jun2021, and we need one Libor fixed on the event date for a loan starting two business days later on 03Jun2021, with coupon paid on 03Sep2021, as well as one discount factor for the payment date. We humans know that from reading the script. A pre-processor figures that out by *visiting* the script at processing time. This allows not only to pre-allocate the vectors of Libors and discounts but also to transform the payoff into something like:

```
0.25 * max( 0, scen[0].libors[0] - STRIKE)
       * scen[0].disc[0] / scen.numeraire
```

The pre-processor also "knows" that event date 0 is 01Jun2021, and that maturity 0 is 03Sep2021 for both Libors and discounts. That information may be communicated to the model before simulations start. The model then knows what it has to simulate and where in pre-allocated memory it must write the simulated data. The evaluator doesn't know what that simulated data is; in particular, it doesn't know what are the maturities of the Libors and discounts. All it does is execute the expression:

```
caplet +=
0.25 * max( 0, scen[0].libors[0] - STRIKE)
       * scen[0].disc[0] / scen.numeraire
```

[7]The model also generates a value for the numeraire under which it performs the simulation, although that numeraire is a scalar value (per event date) that must be simulated and communicated in all circumstances.

while reading the simulated data directly in working memory. All the expensive accounting and look-up, allocation, and matching maturities to working memory (what we call *indexing*) were moved to processing time so that only fast arithmetic operations are performed at simulation time, with direct memory access.

Pre-processing is not limited to indexing simulated data. All variables involved in a script are also pre-indexed so they are random accessed in memory at simulation time. A statement like

```
product pays vAlive * vPayoff
```

is pre-processed into something like

```
V[2] += (V[0] * V[1]) / scen.numeraire
```

where the variables are random accessed at run time in some type of array V, pre-allocated at pre-processing time, where variables are counted and their names are matched to an index in V. Variable indexing is explained in words and code in section 3.3.

Pre-processing is critical to performance. Indexing and other pre-processing steps enable the evaluation of scripts with speed similar to hard-coded payoffs. Pre-processing is facilitated by the framework we develop in part I, with the parsing of scripts into *expression trees* and the implementation of *visitor* objects that traverse the expression trees, gathering information, performing actions and maybe modifying scripts, all while maintaining their own internal state.

1.5 VISITORS

Valuation and pre-processing are two ways in which we *visit* scripts. Other types of visits include queries, which provide information related to the cash flows, for instance the identification of non-linearities; or transformations, like the aggregation and compression of schedules of cash-flow; or decorations, which complement the description of the cash-flows with the payoff of some value adjustment, as explained in part V. And there are many, many others. There is a visitor for everything.

Visitors are all the objects, like the evaluator and the pre-processors, that traverse scripts and conduct calculations or actions when visiting its different pieces, while maintaining an internal state. Their internal state is what makes visitors so powerful. It is through their state that visitors accumulate and process information while traversing scripts. Internal state does not mean that visitors cannot be invoked concurrently. In fact, parallel scripting is easily implemented with multiple visitor instances working in parallel in multiple concurrent threads. The scripting library is thread safe as long as common-sense rules are respected; for example, do not perform parallel work with the same instance of a visitor class.

To make visits possible, we make our scripts *visitable*. This means that we parse them into a data structure that is designed to be traversed in flexible ways. That data structure is the *expression tree* discussed in detail, in words and code, in chapter 2. Trees and visitors are the main building blocks of our scripting library.

One key benefit of the visitor pattern is that it facilitates the support of processes that we haven't even thought of yet. Our visitor-based design caters for needs that will emerge in the future, and enables future development.

For example, we in 2016 worked out an algorithm to automatically smooth all the discontinuities in a transaction in order to stabilize its risk management. This algorithm is based on fuzzy logic and described in detail in part IV. It so happens that an implementation of this algorithm requires the determination of the value domain of all the conditional cash-flows, which means, for every condition of the type $e > 0$, $e \geq 0$ or $e = 0$[8] involved in a script, what is the set of all possible values for e at that point? We designed a visitor to figure that out: the DomainProcessor in scriptingDomainProc.h.

As another example, we were able to significantly improve the performance of LSM regressions for large xVA calculations by pre-computing the entire dependency graph of the many thousands of variables involved in the aggregated script for a large netting set. That allowed us to selectively evaluate, during pre-simulations, only those events that affected the target variable, saving many unnecessary evaluations. And it was relatively straightforward to design a visitor to produce that complete dependency graph.

The visitor-based design provides a framework for the seamless development of any kind of visitor, now and in the future. The visitor-based class designed in section 3.1 takes care of the traversal and visit logic so that a concrete visitor is developed simply by specifying what it does when it visits different nodes in the tree. For instance, the variable indexer developed in section 3.3 takes less than ten lines of code despite the apparent complexity of its work. This visitor effectively counts the variables in a script and matches them to an index in an array.

---

[8]Any condition can be evidently expressed under one of these forms; for example, $e_1 < e_2$ is obviously the same as $e > 0$ with $e = e_2 - e_1$. We built a pre-processor to transform all conditions under these canonical forms.

1.6 MODERN IMPLEMENTATION IN C++

Our library, discussed in part I and provided in our source repository, is designed to facilitate the visit of scripted cash-flows with a convenient internal representation of scripts and a framework for visiting these representations. It revolves around two concepts:

1. *Expression trees* are our internal representation of the scripts in visitable data structures, produced from text scripts through parsing.
2. *Visitors* are the objects that traverse the expression trees, maintaining their own state during traversal, extracting information, performing calculations, or even modifying the trees along the way.

Expression trees are covered in chapter 2. Visitors are covered in chapter 3, including the evaluator in section 3.6 and some important pre-processors in 3.3 and 3.4. Parsing, which turns scripts into expression trees, is covered in the appendix to part I.

We develop our scripting library in self-contained C++. Other implementations attempted to reuse existing programming languages, like Python, Visual Basic, or C#, to code payoffs instead. In this context, the model, typically written in C++, would generate scenarios and delegate the evaluation of payoffs in given scenarios to code written by users in a simpler language. This is attractive at first sight: it saves developers the trouble of implementing a scripting library, and users the trouble of learning it. It offers the power and versatility of a general purpose programming language for the computation of payoffs, providing maximum flexibility in return for a somewhat increased operational risk. The main problem, however, is that it only works for valuation. Python scripts don't describe cash flows; they evaluate payoffs. From the point of view of the C++ code, they are black box functions that can only be executed, not visited. They cannot be pre-processed, queried, or transformed. Such a framework may have merit for the risk management of exotics (if performance and operational risk are not an issue), but it is not suitable for our purpose. This is actually the exact opposite of our design, whose purpose is to provide a representation of cash-flows that is transparent to all kinds of visitors across a system.

We also made the choice to develop a self-contained implementation in standard C++11. We find it unnecessary to recourse to third-party libraries to parse, represent, or visit scripts. Self-contained code offers better control over the algorithms and makes it easier to maintain and extend the code without having to rely on another party. Further, automatic adjoint differentiation (AAD), a technique used to obtain derivative sensitivities with amazing speed, is best implemented when the source code is available. This in itself is reason enough to refrain from using third-party black boxes.

Furthermore, once the fundamental designs and algorithms are well understood, it is natural and relatively painless to produce elegant, efficient, self-contained C++ code, especially with the modern facilities offered by C++11. In particular, we demonstrate in the appendix to section I that the parser (part of the code that turns text scripts into expression trees) is implemented in standard C++ without major difficulty or the recourse to a third-party library. The source code is provided in our repository.

AAD is covered in detail in our publication [27], together with professional C++ code.

1.7 SCRIPT TEMPLATES

Scripting languages are designed to facilitate the description of any derivatives transaction, and, indeed, all transactions must be scripted to benefit from a common representation within a derivatives system. That means that the pricing, booking, and otherwise manipulation of every transaction, down to vanilla swaps and European options, is conducted with scripts. Traders typically manipulate hundreds to thousands of vanilla swaps or European options every day. Although modern scripting is accessible and user-friendly, it is still much easier, faster, and less error prone to fill a few fields on a form, like start and end dates, and a fixed coupon for a swap, than to write a script like:

STARTDATE	01Jun2020
ENDDATE	01Jun2030
FLFREQ	3m
FLBASIS	act/360
FLIDX	L3M
FIXFREQ	1y
FIXBASIS	30/360
CPN	2%
Start: STARTDATE End: ENDDATE Freq: FLFREQ Fixing: start-2bd	swap pays -libor(StartPeriod, EndPeriod, FLBASIS, FLIDX) * cvg(StartPeriod, EndPeriod, FLBASIS) on EndPeriod
Start: STARTDATE End: ENDDATE Freq: FIXFREQ Fixing: start-2bd	swap pays CPN * cvg(StartPeriod, EndPeriod, FIXBASIS) on EndPeriod

For those frequently traded products, it is practical to use *templates*, implemented on spreadsheets or higher level languages, that generate the script automatically out of a small number of inputs.

Templates are typically used for vanilla swaps, European options or first-generation exotics (short-term forex barrier options). They combine the practicality of hard-coded transactions with the benefits of scripting. Many transactions are also regarded as "almost" vanilla swaps, "almost" European, or "almost" standard barriers. In these cases, it is most practical to start with a template and modify the auto-generated script by hand. Templates are also useful when some transactions are booked in a separate system, which is often the case for, say, vanilla swaps. To aggregate these transactions with the rest of a netting set for the calculation of an xVA, for example, those transactions can be scripted on the fly with template code.

This logic may be pushed a step higher, with the design of GUIs, programmed in a language like C#, so that finance professionals without programming expertise, or customers, may build transactions by assembling blocks on a graphical representation of the transaction. This graphical representation is then automatically turned into the corresponding script.

Writing templates is trivial and a template can be implemented directly in a spreadsheet without code. To write a GUI platform for the design of transactions and translate that graphical representation into a script is perhaps more advanced. It certainly requires an expertise in GUI design, something out of the scope of this publication but covered in many recent C# or Python textbooks.

CHAPTER 2

Expression Trees

A scripted product consists of a number of event dates together with the corresponding events, initially represented with text. Text is not convenient for evaluation (or any kind of visit), so our first step is to turn them into appropriate data structures. The process that turns text into our internal representation of events is called *parsing* and covered in the appendix. Our internal representation of an event is a sequence of *expression trees*. We have one expression tree per *statement*.

In our autocallable example, today's event

```
vRef = spot()
vAlive = 1
```

has two statements. We represent each one as a separate expression tree.

A script is a sequence of events, an event is a sequence of statements, and each statement is represented by exactly one expression tree. Visitors visit all the trees of a script in sequence, accumulating information from the previously visited trees that they reuse when visiting subsequent trees.

We define and illustrate expression trees in what follows, first, in words, then, in code.

2.1 IN THEORY

Anatomy of an expression

An expression represents a calculation or an action and consists of a collection of items that represent the elementary operations involved in that calculation or action. An expression that represents an action is called a *statement*. A statement is like a full sentence. An expression that is not a statement is a piece of sentence missing a verb. An event is a sequence of complete sentences, hence, statements.

For example, the statement that described the payoff of a call option:

opt = max(0, spot() – 100)

collects the seven items *opt*, =, *max*, 0, *spot*, –, and 100 into a statement that represents the assignment of the (sub-)expression "max(0, spot() – 100)" into the (sub-)expression "opt":

It transpires from there that an expression is really a hierarchy of (sub-)expressions and a fundamentally recursive thing. An expression is identified by its higher-level item, = in our example, the operation evaluated last in the expression, which depends on the other items, without itself being an argument to any other operation. The assignment operation has two arguments: the assigned variable, "opt" in our example, and the assigned amount, "max(0, Spot() – 100)." The left-hand side (LHS) "opt" is an expression that consists of the single item *opt*, which represents the assigned variable. The right-hand side (RHS) "max(0, Spot() – 100)" is itself an expression (but not a statement), which represents the assigned amount and is identified by its higher level item *max*:

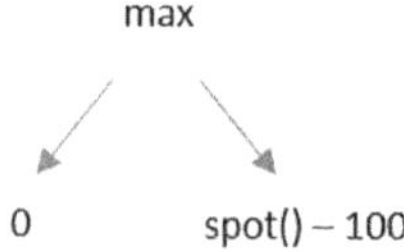

The binary *max* operation depends on two (sub-)expressions: the constant "0" (constant item 0) and the (sub-)expression "Spot() – 100," which is a subtraction expression identified by its – top level item, itself depending on its LHS "Spot()" (the item *spot*() representing the underlying asset's simulated value) and its RHS "100" (the constant item 100), both of which are single item expressions without dependencies, hence, ending the recursion:

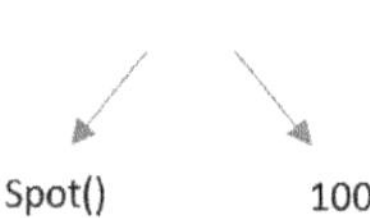

We broke the entire statement down into a hierarchical representation of its items, where each item represents an elementary calculation or action, which may depend on a number of arguments, which are themselves expressions broken down into a hierarchical representation of its items, and so forth, down to special, self-contained items that represent operations without arguments:

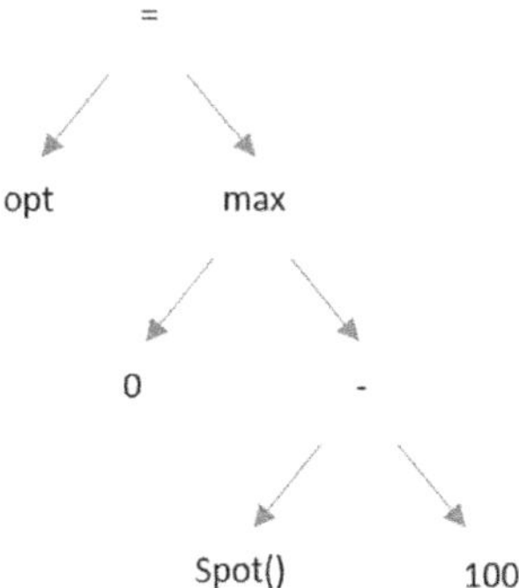

A recursive definition

This recursive, hierarchical representation of the items in an expression, which we just performed manually, is called an *expression tree*. The process of establishing the tree for an expression is called *parsing*. We just parsed a statement manually, as a pedagogical exercise, although parsing, thankfully, can be fully automated. Our appendix explains the parsing process in words and code.

The items in an expression tree are called *nodes* and represent elementary operations, those "atomic" operations that cannot be further broken down. Expression trees are always identified by their top node. The arguments to an operation are themselves expression trees, identified by their own top nodes, which we call the *child nodes* to the parent operation. Operations without arguments like constants or variables are called *leaves*. The definition of an expression tree is therefore a recursive one. An expression tree is identified by a top node. A node represents an operation and may have children, themselves expression trees identified by their top nodes. Expression trees without children are leaves. For now, we have three types of leaves: constants, variables, and simulated data.

Readers familiar with GOF design patterns recognized the so-called *composite* pattern where a whole is a collection of parts all identical in nature (in the sense of a virtual base class) to the whole, see [12]. In the composite pattern, it is the nature of a node (in the sense of a concrete class), not the data it stores, that represents a particular operation. The top node in our statement *is* an assignment node. The data it holds are a pair of references to its kids, the top nodes of its LHS and RHS (sub-)trees, stored as base class pointers to the top nodes of those trees, in accordance with the composite pattern. An expression tree is therefore a recursive data structure, and a peculiar one, in that it represents an expression *in its DNA*, and not its contents.

Expression trees are designed so they can be *visited*. For now, let us focus on one particular form of visit, that is, the *execution* of the statement, also called the *evaluation* of the tree, given a scenario for the simulated data.

Evaluate an expression tree means evaluate its top node, in our example, perform an assignment. The assignment cannot be performed before the variable being assigned is identified and the assigned amount computed. Hence, the evaluation of the assign node starts with the evaluation of its child nodes. Only then can the assignment be executed. Hence, the *opt* node is evaluated first so the assigned variable is known, followed by the evaluation of the RHS tree to calculate the assigned amount. In the same, recursive manner, to evaluate that tree means evaluate its top *max* node, which requires, first, an evaluation of its arguments (then, we keep the largest one). The first argument to *max* is a constant leaf that evaluates to 0, while its second argument is another (sub-)tree with a top node −. Hence, the evaluation of = triggers the evaluation of *max*, which triggers the evaluation of −, which triggers the evaluation of *spot*() (which evaluates to the simulated underlying asset price) and the constant 100. The evaluation order is therefore *opt*, 0, *Spot*(), 100, −, *max*, =. The evaluation is fired on the top node, but, because the arguments to an operation evaluate before that operation, the effective evaluation always and automatically proceeds bottom-up, leaves to top.

It is a general property of expression trees that their evaluation, fired on the top node, always proceeds bottom-up. This is an immediate consequence of the rule, hard-coded in the evaluator, and many other visitors, that *the visit of every node begins with a visit to its arguments*. Hence, a visit to the top node automatically, *recursively* visits the whole tree, bottom-up, starting with leaves and ending with the top node. Further, each node is visited exactly once. Such traversal pattern is called *depth first* (because it starts with the leaves), or *postorder* (because kids are visited before their parents).

Notice also, that, in order to evaluate the sequence *opt*, 0, *Spot*(), 100, −, *max*, = into a result given a scenario for the spot, intermediate results must be stored while the tree is being visited. The data structure most appropriate to hold the intermediate results of a depth-first visit is a stack, also called LIFO (Last In First Out) container, that returns its contents in the reverse order from their acquisition. The following figure shows the state of the stack after the evaluation of the nodes in the postorder sequence:

postorder	node	evaluation	number stack	variable stack
0				
1	opt	push (address of) opt on variable stack		opt
2	0	push 0 on number stack	0	opt
3	spot()	push spot on number stack	spot 0	opt
4	100	push 100 on number stack	100 spot 0	opt
5	-	pop 2 top elements from number stack subtract 1st from 2nd push result on number stack	spot-100 0	opt
6	max	pop 2 top elements from number stack push largest on number stack	(spot-100)+	opt
7	=	pop top elements from both stacks assign number to variable		

Examples

These notions are so fundamental to scripting that we illustrate them further with examples, at the risk of repetition, before moving on to the implementation. The first event in our simplified autocallable script consists of two statements, both of which are assignments and map to the following expression trees:

```
vRef = spot()
vAlive = 1
```

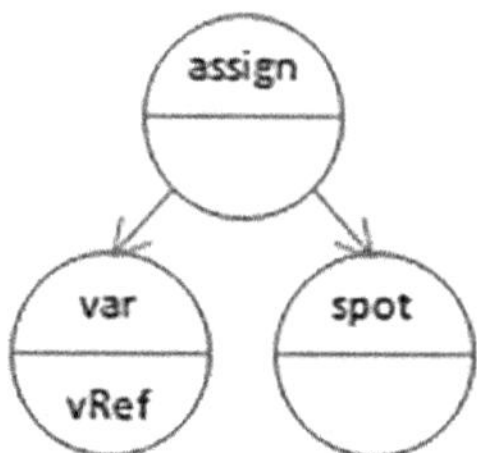

and

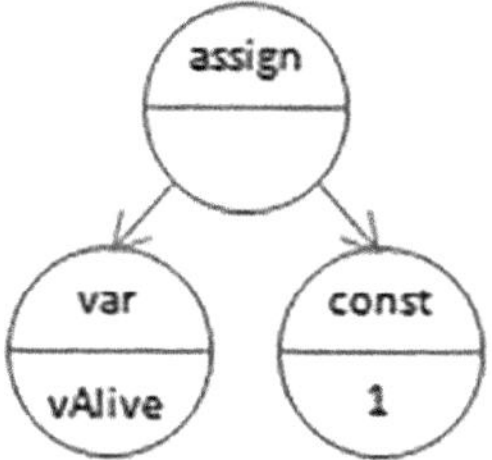

Another, maybe less trivial example like $x = (2 * log(y) + 5) * z$ maps, to the following tree:

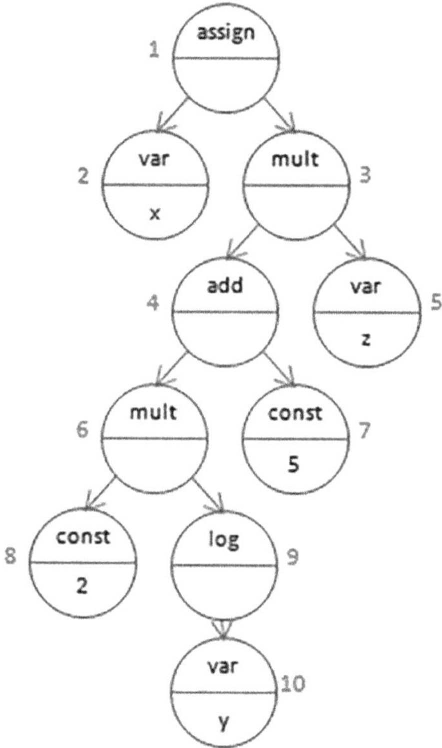

Let us clarify its evaluation. The second argument to the top node 1 is the node 3, top of a sub-tree that represents the assigned expression $(2\log(y)+5)z$. This node 3 is a multiplication node with arguments $2\log(y)+5$ and z. Its first argument is the top node 4 of a sub-tree that represents $2\log(y)+5$. This is an addition node with arguments $2\log(y)$ and 5. Every node is the top of its own sub-tree representing its own sub-expression, where every argument to a node is another expression tree identified by its top node. This is truly a recursive data structure, naturally designed for a recursive traversal.

The second argument to the addition node 4 is the node 7 standing for the constant 5. This is a node without further arguments. It represents an expression that does not need to evaluate other nodes before it evaluates itself (to the constant 5). All branches in a tree end with leaves. Leaves are where the recursion *first evaluate my arguments then evaluate me* stops and the actual evaluation starts.

The evaluation of node 1 first evaluates its arguments 2 and 3. The evaluation of leaf 2 resolves in the assigned variable x. The evaluation of node 3 first evaluates its arguments 4 and 5. The evaluation of node 4 first evaluates its arguments 6 and 7. The evaluation of node 6 first evaluates nodes 8 (which resolves into the constant 2) and 9, which first evaluates the node 10, which resolves in the value of the variable y.

Then node 9 can be evaluated, and then node 6. The evaluation of node 7 resolves in the constant 5, so node 4 can be evaluated, then node 5 reads the value of the variable z, node 3 is evaluated and finally node 1. The order of evaluation is 8-10-9-6-7-4-5-3-2-1. Once the tree is properly constructed, a full evaluation of the whole tree in the correct order, leaves to top, is guaranteed, whenever the top node is evaluated, courtesy of the rule *visit my kids first, then do something with me*. That postorder rule also guarantees that every node is visited exactly once.

An expression tree can be represented graphically (like we did) or *functionally*. Our $x = (2 * log(y) + 5) * z$ tree above can be written in the compact function form:

```
Assign( x , Mult( Add( Mult( 2 , Log( y ) ) , 5 ) , z ) )
```

This representation is less intuitive than the graphical form, but it is also more compact and highlights the evaluation order, from innermost to outermost functions.

Note that we illustrated the evaluation of the tree for clarity, but everything we said applies to all forms of postorder visits.

Carrying on with our autocallable example, the next event is

```
if spot() > vRef then prd = 110 vAlive = 0 endIf
```

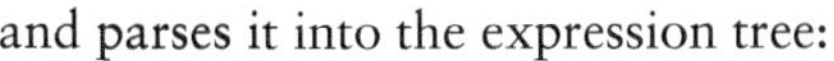

and parses it into the expression tree:

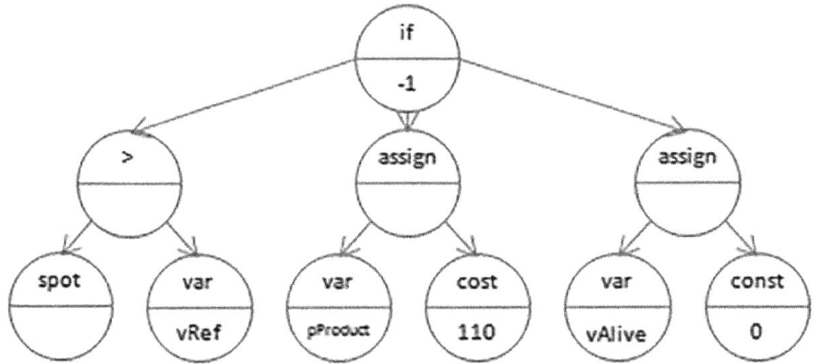

The −1 data in the top *if* node will be explained in a minute. Its first argument (first sub-tree on the left) is the condition, and subsequent arguments on the right are statements to be executed depending on the condition being evaluated to true or false.

Moving to the next event:

```
if spot() > vRef then
if vAlive = 1 then prd = 120 endIf
vAlive = 0
endIf
```

We get:

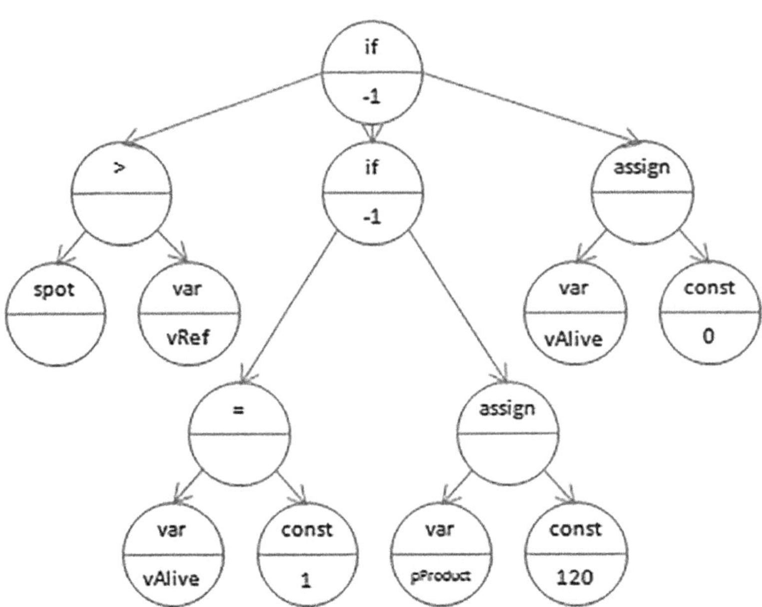

And finally the last event:

```
if vAlive = 1 then
if spot() > vRef then prd = 130
else prd = 50 endIf
endIf
```

gives us:

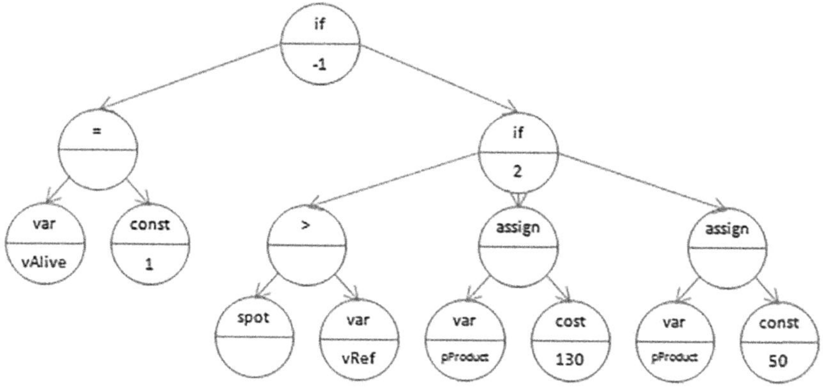

The nested *if* node on level 2 holds data 2 to signify that when the condition is true, arguments 1 up to but not including 2 must be evaluated (condition is argument 0); when the condition is false, arguments 2 and above are to be evaluated. So the node holds the index of the argument corresponding to the first *else statement*. A data of −1 means we have no else.

When a product delivers more than one payment, like a cliquet, that pays the positive performance of some index over multiple periods, we may use the somewhat inelegant but working syntax[1]

```
prod = prod + ...
```

[1]Later we show how we can do much better than that.

Hence, a standard cliquet can be scripted as (today is 1st June 2020):

01Jun2020	vRef = spot()
01Jun2021	vPerf = Max(spot() - vRef, 0) / vRef pCliquet = vPerf vRef = spot()
01Jun2022	vPerf = Max(spot() - vRef, 0) / vRef pCliquet = pCliquet + vPerf vRef = spot()
01Jun2023	vPerf = Max(spot() - vRef, 0) / vRef pCliquet = pCliquet + vPerf

where the repeated statement

```
pCliquet = pCliquet + vPerf
```

has a tree representation:

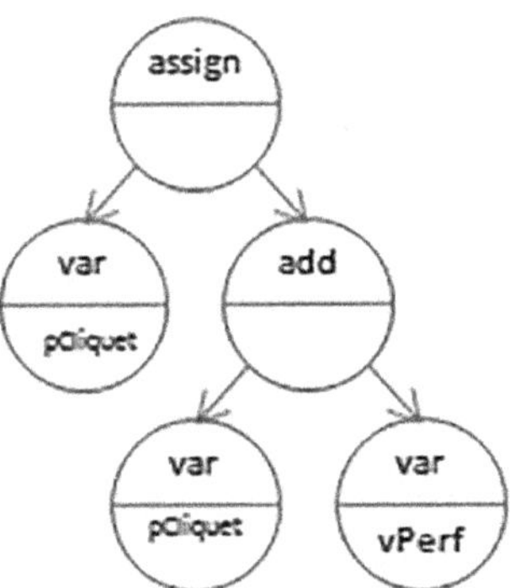

On evaluation, the current value of the variable pCliquet is first read from the bottom left node, added to *vPerf*, and the result is assigned back to *pCliquet*. When evaluation completes, *pCliquet* holds the sum of the payments in a given scenario (path for the spot on the four event dates). We will implement a more expressive, elegant, and practical syntax in chapter 5.

This closes our theoretical exploration of expression trees and we turn to their implementation in C++.

2.2 IN CODE

In this section, we show how data structures for the representation of expression trees are implemented in C++ with the classic object-oriented (OO) *composite* pattern. Although a more efficient implementation is obtained with modern constructs like *tagged unions* (implemented in C++17 with the *variant* class), we present the classic OO implementation in order to convey the key ideas with simple code. Our code may be rewritten with modern constructs without major difficulty after the fundamental notions are well understood.

A product is a collection of events indexed by event dates. We can represent a product as follows:

```
using namespace std;
#include <vector>

// Date class from your date library
// class Date;
using Date = int;

class Product
{
  vector<Date>  myEventDates;
  vector<Event>  myEvents;

public:

  // Access event dates
  const vector<Date>& eventDates()
  {
    return myEventDates;
  }
  // Events are not accessed,
  // remain encapsulated in the product
};
```

An event is an ordered collection of statements, and we have exactly one expression tree per statement:

```
using Statement = ExprTree;
using Event = vector<Statement>;
```

An expression tree is identified by its top node. We hold them by smart pointers so that an expression tree is a (smart) pointer on its top node.

```
#include <memory>
using ExprTree = unique_ptr<Node>;
```

The composite pattern is implemented by making *Node* a hierarchical class that is declined into different types of concrete *Nodes*, one for every operation. All *Nodes* may have arguments that are themselves (sub-)expression trees. Hence, we hold a vector of arguments on the base class:

```
struct Node
{
  vector<ExprTree> arguments; // ExprTree = base pointer on (top) node

  virtual ~Node() {}
};
```

Expression trees are held as (smart) base *Node* pointers referring to their top *Node*, in accordance with the recursive structure. Smart pointers ensure that whenever an *ExprTree* exits scope, the full tree beneath is released from memory. Hence, the top node that identifies a tree also owns the memory of the entire tree. C++11's *unique_ptr*s guarantees zero cost compared to dumb pointers (contrarily to *shared_ptr*s, which would be a poor choice here). The fact that arguments are base class pointers is the whole purpose of the composite pattern, where child nodes are held in a data structure of the same nature as parents, and whole trees, including subtrees, are identified by their top node and held by a base class pointer on that node.

We made nodes a *struct* rather than a *class* because nodes are meant to be visited and visitors need access to the node's data. We could have made all visitors friend classes, but that would require boilerplate code, since friendship is not inherited in C++. We therefore make an exception to C++ best practices and expose the node internals.

Each elementary operation is represented by its own concrete *Node* type. Let us enumerate all the concrete nodes. We have the unary +/−:

```
struct NodeUplus : public Node {};
struct NodeUminus : public Node {};
```

The math operators:

```
struct NodeAdd : public Node {};
struct NodeSubtract : public Node {};
struct NodeMult : public Node {};
struct NodeDiv : public Node {};
struct NodePow : public Node {};
```

The math and financial functions like:

```
struct NodeLog : public Node {};
struct NodeSqrt : public Node {};
struct NodeMax : public Node {};
struct NodeMin : public Node {};
// ...
```

The comparators and condition operators:

```
struct NodeEqual : public Node {};
struct NodeDifferent : public Node {};
struct NodeSuperior : public Node {};
struct NodeSupEqual : public Node {};
struct NodeInferior : public Node {};
struct NodeInfEqual : public Node {};

struct NodeAnd : public Node {};
struct NodeOr : public Node {};
```

The assignment node:

```
struct NodeAssign : public Node {};
```

The *spot* node for the access of the underlying asset price:

```
struct NodeSpot : public Node {};
```

The *if* node:

```
struct NodeIf : public Node
{
  int   firstElse;
};
```

The *const* node for holding numerical constants:

```
struct NodeConst : public Node
{
   NodeConst(const double v) : val(v) {}

  double val;
};
```

And finally the *var* node for variables:

```
struct NodeVar : public Node
{
   NodeVar(const string n) : name(n) {}

  const string name;
};
```

Notice that the concrete node types store no data with the exception of *if* (which holds the index of first else statement or −1), *const* (holds its value), and *var* (holds the name of the variable). It is the structure of the tree and the type of its nodes that represent the statement and specify it entirely. Notice also the absence of methods. All the action happens in the visitors, not in the tree. It follows from all this that the trees are immutable by evaluation, something of considerable significance for parallel simulations (see our dedicated publication [27]).

For convenience, we also produce two factory functions for building nodes. The first one produces a base *Node* (smart) pointer, that is an *ExprTree*, and the second one produces a concrete *Node* pointer of a specified type. These factories help parsing, and avoid the direct application of the operator *new*, shielding us from memory leaks. We (purposely) derived the constant and variable nodes from a default constructor, and our factory functions take variadic parameters and forward them to constructors, so they may be used with any node type, with or without constructor arguments, in accordance with a now classic C++11 pattern:

```
template <typename NodeType, typename... Args>
unique_ptr<NodeType> make_node(Args&&... args)
{
  return unique_ptr<NodeType>( new NodeType(forward<Args>(args)...));
}

template <typename NodeType, typename... Args>
unique_ptr<Node> make_base_node(Args&&... args)
{
  return unique_ptr<Node>( new NodeType(forward<Args>(args)...));
}
```

The transformation of the events from strings into collections of expression trees is performed by the *parsing* algorithm. We use a version of the recursive descent parser, an elegant, well-known parsing algorithm with linear complexity. It implements the parse function in:

```
class Product
{
  vector<Date>  myEventDates;
  vector<Event>  myEvents;

public:

  // ...

  // Build events out of strings

  template<class EvtIt>
  // Takes begin and end iterators on pairs of
  // dates and corresponding event strings
  //  as from a map<Date,string>

  void parseEvents( EvtIt begin, EvtIt end)
  {
    // Copy event dates and parse event strings sequentially
    for( EvtIt evtIt = begin; evtIt != end; ++evtIt)
    {
      // Copy event date
      myEventDates.push_back( evtIt->first);

      // Parse event string
      myEvents.push_back( parse( evtIt->second));
    }
  }
};
```

The implementation of the function *parse()* is detailed and discussed in the appendix.

After parsing, our *Product* holds a vector *myEvents* of events corresponding to the vector *myEventDates* of event dates. Each event is a collection of statements; each statement is an expression tree. The product is ready for *visits*.

The expression trees hold in their DNA the operations that build the cash-flows. This information is ordered by dependency and ideally suited for analysis, modification, or evaluation from various parts of the system. We call all these *visits*, and the components that perform them, *visitors*.

We advise readers unfamiliar with parsing to jump to the appendix and learn how exactly we build expression trees from text, before moving on to visitors.

CHAPTER 3

Visitors

3.1 THE VISITOR PATTERN

Visitors are objects that manipulate expression trees. Visitors traverse the trees in a given order (often depth-first postorder) to extract information and/or perform actions, while maintaining internal state. Once again, in order to convey the principal ideas with simple code, we present the classic OO visitor pattern based on double virtual dispatch: what code is executed when a visitor visits a node in the tree depends on the concrete class of the visitor and the concrete class of the node. A more efficient implementation could lift virtual dispatch overhead. As commented previously, nodes could be implemented with tagged unions in place of a virtual hierarchy. Visitors also do not have to constitute a virtual hierarchy; they could be implemented as template arguments in general visitation procedures. The C++17 variant construct has built-in support for visitors. In practice, those optimizations make code harder to understand for readers unfamiliar with modern C++ constructs, and the improvement in performance is rather limited. Readers who wish to modernize our code will not meet significant difficulties once the central ideas are well understood.

Let us warm up with a particularly easy visitor, the *Debugger*, who prints out a parsed expression tree. Once parsing is implemented, we may want to throw hundreds of event strings at it and check that they parse correctly. Unfortunately, to check the structure of trees in a debugger is less than ideal. Debuggers are not built to easily visualize composite patterns. Therefore, we want some functionality to traverse trees bottom-up and write their *functional* form. We want a tree like

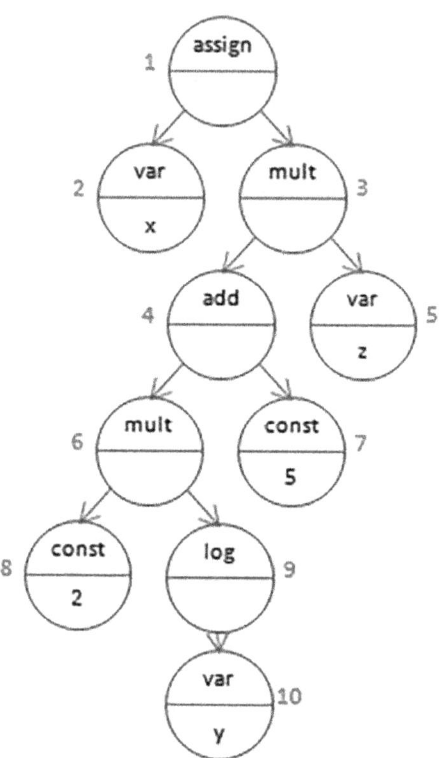

to come out as:

ASSIGN(VAR[X] , MULT(ADD(MULT(CONST[2] , LOG(VAR[Y])) , CONST[5]) , VAR[Z]))

This may not be particularly friendly but good enough for debugging our parser. Readers with GUI expertise may code a software that turns a functional form into the picture of a tree.

We need an object, let us call it *Debugger*, that traverses the tree, visiting each node in the evaluation sequence 8-10-9-6-7-4-5-3-2-1, to recursively write its functional form. We remind that the order of evaluation is guaranteed as long as the visit to every node starts with the visit to its arguments. In this case, a visit to the top node traverses the whole tree in the correct order automatically.

When the *Debugger* hits the *add* node number 4, for example, it first visits its arguments 6 and 7, stores their functional form (let's call them *F6* and *F7*), and since our node is of type *NodeAdd*, it computes and stores its function form $F4 =$ "*Add(F6, F7)*." When the debugger visits the *mult* node 6, however, it computes and stores its functional form $F6 =$ "*Mult(F8, F9)*." When it visits the variable leaf 10, it writes its name y: $F10 =$ "*Var*[y]." Our debugger does different things according to the concrete type of the visited node. The only thing it always does irrespective of the node is to visit its arguments first.

Our debugger is the first in a long series of visitors. Other visitors do different things with the visited nodes. In particular, the *evaluator*, the visitor that evaluates trees against scenarios, when visiting node 4, also visits arguments 6 and 7 first (evaluates them numerically and stores the results), but then it evaluates and stores the result for node 4 as a number, not a string, contrary to the debugger: for the evaluator, $R4 = R6 + R7$. When it visits node 6, it computes $R6 = R8 * R9$. And on the leaf 10, it stores into $R10$ the value of the variable y, not its name. The evaluator acts on nodes differently to the debugger, the only common action being to visit arguments first to ensure the whole tree is traversed in the correct order.

So, what happens when a visitor hits a node depends on the concrete types of both the node and the visitor. C++ classically provides virtual method overriding as a native mechanism for subordinating an action to the concrete type of *one* object. C++ does not natively provide a mechanism for the subordination of a task to the concrete types of *two* objects, also called *double dispatch*. In order to overcome this limitation, GOF produced the influential *visitor* design pattern, which is explained next in the context of our expression trees. We refer to GOF (see [12]) or one of the numerous textbooks on design patterns for a more general discussion.

First, we need a *Visitor* base class with virtual functions for visiting the different concrete node types. Instead of pure virtual functions, we default all visits to traversing the node into its arguments. This way, visitors don't need to override the visit methods for node types that are not of interest to them (for instance, a visitor that only looks into variables only needs to override the method for visiting variables), the default node traversal method ensuring that the whole tree is still traversed in the correct order.

```
class Visitor
{
protected:
  // Protected constructor so the base class cannot be instantiated
  Visitor() {}

public:
  virtual ~Visitor() {}

protected:
  // Default visit just visits arguments so as to ensure that
  // the whole sub-tree is visited
  virtual void visitArguments( Node& node)
```

```
  {
    for( auto& arg : node.arguments) arg->acceptVisitor( *this);
  }

public:

  // Entry point for visiting a (sub-) tree
  void visit( ExprTree& tree)
  {
    tree->acceptVisitor( *this);
  }

  // All concrete node default visitors,
  // visit arguments unless overridden

  virtual void visitUplus( NodeUplus& node)
  { visitArguments( node); }
  virtual void visitUminus( NodeUminus& node)
  { visitArguments( node); }
  virtual void visitAdd( NodeAdd& node)
  { visitArguments( node); }
  virtual void visitSubtract( NodeSubtract& node)
  { visitArguments( node); }
  virtual void visitMult( NodeMult& node)
  { visitArguments( node); }
  virtual void visitDiv( NodeDiv& node)
  { visitArguments( node); }
  virtual void visitPow( NodePow& node)
  { visitArguments( node); }
  virtual void visitLog( NodeLog& node)
  { visitArguments( node); }
  virtual void visitSqrt( NodeSqrt& node)
  { visitArguments( node); }
  virtual void visitMax( NodeMax& node)
  { visitArguments( node); }
  virtual void visitMin( NodeMin& node)
  { visitArguments( node); }
  virtual void visitEqual( NodeEqual& node)
  { visitArguments( node); }
  virtual void visitDifferent( NodeDifferent& node)
  { visitArguments( node); }
  virtual void visitSuperior( NodeSuperior& node)
  { visitArguments( node); }
  virtual void visitSupEqual( NodeSupEqual& node)
  { visitArguments( node); }
  virtual void visitInferior( NodeInferior& node)
  { visitArguments( node); }
  virtual void visitInfEqual( NodeInfEqual& node)
  { visitArguments( node); }
  virtual void visitAnd( NodeAnd& node)
  { visitArguments( node);}
  virtual void visitOr( NodeOr& node)
  { visitArguments( node); }
  virtual void visitAssign( NodeAssign& node)
  { visitArguments( node); }
```

```
  virtual void visitPays( NodePays& node)
  { visitArguments( node); }
  virtual void visitSpot( NodeSpot& node)
  { visitArguments( node); }
  virtual void visitIf( NodeIf& node)
  { visitArguments( node); }
  virtual void visitConst( NodeConst& node)
  { visitArguments( node); }
  virtual void visitVar( NodeVar& node)
  { visitArguments( node); }
};
```

Note that all visit methods for concrete node types must be explicitly written even though they all do the exact same thing. C++ does not support virtual template methods, so there is unfortunately no simple or elegant workaround.

We also conveniently code loop-over events and statements in our Product class for visiting all the events in a product sequentially:

```
class Product
{
  vector<Date>   myEventDates;
  vector<Event>  myEvents;

public:

  // ...

  // Sequentially visit all statements in all events
  void visit( Visitor& v)
  {
    // Loop over events
    for( auto& evt : myEvents)
    {
      // Loop over statements in event
      for( auto& stat : evt)
      {
        // Visit statement
        v.visit( stat);
      }
    }
  }
};
```

The entry point for the visit of a tree in the base *Visitor* class calls the method *acceptVisitor*() on the top node, passing a reference to this visitor as an argument. The default traversal method visits arguments in the same way, calling *acceptVisitor*() on each of the arguments. The reference to $*$ *this* carries the concrete type of the visitor and it is passed to the *acceptVisitor*() method on the node, which also knows its own concrete type. For this syntax to work, we must have a virtual *acceptVisitor*() method on the base *Node* object, and override it for all concrete *Node* types.

```
struct Node
{
  vector<ExprTree> arguments;

  virtual ~Node() {}

  virtual void acceptVisitor( Visitor& visitor) = 0;
};

// Unary +
struct NodeUplus : public Node
{
  void acceptVisitor( Visitor& visitor) override;
};

void NodeUplus::acceptVisitor( Visitor& visitor)
{
  visitor.visitUplus( *this);
}

// Unary -
struct NodeUminus : public Node
{
  void acceptVisitor( Visitor& visitor) override;
};

void NodeUminus::acceptVisitor( Visitor& visitor)
{
  visitor.visitUminus( *this);
}

// operator +
struct NodeAdd : public Node
{
  void acceptVisitor( Visitor& visitor) override;
};

void NodeAdd::acceptVisitor( Visitor& visitor)
{
  visitor.visitAdd( *this);
}

// Etc
```

In summary, visitors visit nodes by calling their virtual *acceptVisitor* method on a base *Node* class reference, and passing a base *Visitor* reference to themselves. The node calls the overridden *acceptVisitor*() for the concrete *Node* type, which in turn calls the *visitSomething* virtual function on the base *Visitor* reference, which ends up calling the overridden *visitSomething*() on the concrete *Visitor* type. If this particular visitor overrides its visit for *something*, *visitSomething*(), then the overridden method is called. Otherwise, the base *visitSomething*() traverses the arguments. *visitSomething*() is passed a reference to the concrete *something* node, so it can access information in that *Node*, including arguments. These mechanics effectively implement double dispatch using overriding twice, once on the node and once on the visitor.

We illustrate the pattern with the development of a simple concrete visitor, the *Debugger*.

3.2 THE DEBUGGER VISITOR

Every visitor that does something with a given node type other than just traverse the node to visit its arguments, must override its visiting method for this particular node type. Our *Debugger* overrides the visit of the different types of nodes into:

```
void visitUplus( NodeUplus& node) override
    { debug( node, "UPLUS"); }
void visitUminus( NodeUminus& node) override
    { debug( node, "UMINUS"); }
void visitAdd( NodeAdd& node) override
    { debug( node, "ADD"); }
// ...
void visitSpot( NodeSpot& node) override { debug( node, "SPOT"); }
void visitIf( NodeIf& node) override
{
    debug(node, string( "IF[")+to_string( node.firstElse)+']');
}
void visitConst( NodeConst& node) override
{
    debug(node, string( "CONST[")+to_string( node.val)+']');
}
void visitVar( NodeVar& node) override
{
    debug( node,
        string( "VAR[")+node.name+','+to_string( node.index)+']');
}
```

All nodes but *if*, *const*, and *var* nodes just call the debugger's *debug*() method with a string that identifies the type of the node. We will come back to this *debug*() method in an instant. *if*, *const*, and *var* nodes also call *debug*(), but pass their internal data in addition to their type so we can see that, too.

Let's see double dispatch at work here. When the debugger visits some node, say an *add* node, it calls the virtual *acceptVisitor*() method on the node, passing a reference to itself as an argument. This call resolves through classical overriding to a call to the *NodeAdd*'s overridden *acceptVisitor*():

```
void NodeAdd::acceptVisitor( Visitor& visitor)
{
    visitor.visitAdd( *this);
}
```

This in turn makes a call to the overridden *visitAdd*() method of our concrete visitor, in this case the *Debugger*, again through classical overriding, since the base *Visitor* class reference passed as an argument to *acceptVisitor*() carries its concrete type. So in the end, when the debugger hits an add node, the *visitAdd*() method on the debugger is called with a reference to the add node as an argument. Exactly what we want. This is the visitor design pattern. It may seem convoluted at first, but once we wrap our mind around it, we realize that it effectively implements the double dispatch mechanics missing in C++.

Our debugger is complete once we have its debug method.

```
class Debugger : public Visitor
{
  string    myPrefix;
  quickStack<string>  myStack;

  // The main function call from every node visitor
  void debug( const Node& node, const string& nodeId)
  {
    // One more tab
    myPrefix += '\t';

    // Visit arguments, right to left
    for( auto it = node.arguments.rbegin();
        it != node.arguments.rend(); ++it)
          (*it)->acceptVisitor( *this);

    // One less tab
    myPrefix.pop_back();

    string str( myPrefix + nodeId);
    if( ! node.arguments.empty())
    {
      str += "(\n";

      // First argument, pushed last
      str += myStack.top();
      myStack.pop();
      if( node.arguments.size() > 1) str += myPrefix + ",\n";

      // Args 2 to n-1
      for( size_t i=1; i<node.arguments.size()-1; ++i)
      {
        str += myStack.top() + myPrefix + ",\n";
        myStack.pop();
      }

      if( node.arguments.size() > 1)
      {
        // Last argument, pushed first
        str += myStack.top();
        myStack.pop();
      }
```

```
      // Close ')'
      str += myPrefix + ')';
    }

    str += '\n';
    myStack.push( move( str));
  }

public:

  // Access the top of the stack,
  // contains the functional form after the tree is traversed
  string getString() const
  {
    return myStack.top();
  }

  // All concrete node visitors,
  // visit arguments by default unless overridden
  void visitUplus( NodeUplus& node) override { debug( node, "UPLUS"); }
  // ...
  void visitVar( NodeVar& node) override
  {
    debug( node,
        string( "VAR[")+node.name+','+to_string( node.index)+']');
  }
};
```

The debugger uses a stack of strings to store intermediate results while traversing trees.[1] When it visits a node, it starts with a visit to the arguments, right to left, so that we later pop the results from the stack left to right. The visit is always performed with a call to the virtual *acceptVisitor*() on the visited node, passing a reference to this visitor for argument. This guarantees a correct double dispatch, and that the tree is traversed in full in the correct order. All that remains is to write the name of the node, pop arguments from the stack, write them inside parentheses and separated by commas after the node name, and push the result onto the stack. We add some formatting logic for parentheses, commas, and tabs. The recursive nature of visits takes care of the rest.

A call to the top node of a tree's *acceptVisitor*() method with a *Debugger* object for argument (or equivalently a call to the debugger's *visit*() method with the top node of the tree as argument) will result in a full tree traversal. When the traversal completes, the stack holds one string with the end result,

[1]The standard STL stack adapter has poor performance, at least with the compilers we tested. Stacks are a core part of parsing and scripting in general; therefore, we use our own simple, fast implementation, available online with the rest of our code. Alternatively, readers may revert to the STL stack with an alias template <class T> using quickStack<T> = stack<T>.

the functional form of the whole tree. We need a function to access the result after a tree is visited, hence, *getString*().

We use the debugger to check our parser with a variety of event strings. We may use the following driver:

```
void parsingTester()
{
  string evtStr;

  cout << ">> ";
  getline( cin, evtStr);

  while( evtStr != "quit")
  {
    try
    {
      vector<ExprTree> trees = parse( evtStr);

      for ( auto& tree : trees)
      {
        Debugger debug;

        tree->acceptVisitor( debug);

        cout << debug.getString() << endl;
      }

    }
    catch( const script_error& e)
    {
      cout << e.what() << endl;
    }

    cout << ">> ";
    getline( cin, evtStr);
  }
}
```

The debugger is essentially a play tool for us developers; however, we encourage readers to take advantage of its simplicity to wrap their minds around the recursive nature of its visits and the use of a stack to store intermediate results while visiting trees depth first. More sophisticated visitors are built around the same logic, and it helps to understand it first in the context of a simple one.

3.3 THE VARIABLE INDEXER

Before we move on to the evaluator, the visitor that evaluates expression trees at run time, we prepare the trees for an efficient evaluation. We call this process *pre-processing*, and we call *pre-processors* the visitors who perform it. One of such visitors is the *variable indexer*.

When a statement like $x = y$ is executed during simulations, the evaluator must look up a variable named y and read its value, then look up the address of another variable named x, and write the value of y into it. These lookups occur on every statement with variables, in all the simulated scenarios, causing an unacceptable performance drain. We expect valuation of scripted products to match the performance of efficient hard-coded pricers. This is not going to happen if we look up variable names repeatedly at run time. We want the evaluator to pick directly the values and addresses of variables in memory. Hence, variables must be pre-indexed, so that if, say, x takes index 3 and y takes index 7, $x = y$ is read by the evaluator as $var[3] = var[7]$ with a direct, random memory access.

To perform this indexing before simulations take place, we use a visitor that we call variable indexer. This particular visitor traverses all the trees like all the visitors, but this one stops only to visit variable nodes. While visiting variables, it counts them and sets their unique index on their node. Not only is indexing a strong performance requirement, it also highlights some very desirable features of visitors that did not appear that clearly in the debugger:

1. Visitors only override visiting methods for the nodes they are interested in, in this case, variable nodes. The whole trees will still be traversed, courtesy of the default virtual visit methods in the base visitor class. Hence, some visitors' code may be quite compact.
2. Visitors hold a state that they carry over as they visit sequences of trees and iterate through statements and events, updating their state according to the nodes they visit, and using the information gathered on past nodes as they visit the current ones. In this case, our indexer keeps a map of all the variables it encountered.
3. Visitors are not limited to reading information stored in expression trees. The variable indexer writes the index of variables in the variable nodes. Other visitors may even modify the structure of the trees.

First, we need to add an index on the variable nodes:

```
struct NodeVar : public Node
{
  string   name;
  unsigned  index;

  void acceptVisitor( Visitor& visitor) override;
};
```

Then, we develop the visitor. This particular one stops only for variables, so it overrides only the *visitVar*() method, making its code very compact. Furthermore, it carries a state coming from all the variables it has visited since its construction as a map from variable names to their index.

```
#include <map>

class VarIndexer : public Visitor
{
  // State
  map<string, unsigned> myVarMap;

public:

  // Variable indexer: build map of names to indices
  // and write indices on variable nodes
  void visitVar( NodeVar& node) override
  {
    auto varIt = myVarMap.find( node.name);
    if( varIt == myVarMap.end())
      node.index = myVarMap[node.name] = myVarMap.size();
    else node.index = varIt->second;
  }
};
```

That's it. When it visits a variable node, the indexer looks up its name in its map. If found, it sets the corresponding index on the node. If not, it adds the name on its map for future reference, with the next available index, that is, the current size of the map, and sets the same index on the node. The whole code takes three lines and this is all it takes for the visitor to index all variables in a collection of trees across statements and events. We need to add a method to produce the resulting array of variable names:

```
class VarIndexer : public Visitor
{
  // State
  map<string,unsigned> myVarMap;

public:

  // Access vector of variable names
  // v[index]=name after visit to all events
  vector<string> getVarNames() const
  {
    vector<string> v( myVarMap.size());
    for( auto varMapIt = myVarMap.begin();
        varMapIt != myVarMap.end();
        ++varMapIt)
    {
      v[varMapIt->second] = varMapIt->first;
    }

    // C++11: move not copy
    return v;
  }

  // ...
};
```

and add the resulting array of variables as a data member of our *Product* class, so its knows how many variables we have, and what their names are, for retrieval after simulations are complete.

```
class Product
{
  vector<Date>  myEventDates;
  vector<Event>  myEvents;
  vector<string>  myVariables;

public:

  // ...

  // Index all variables
  void indexVariables()
  {
    // Our indexer
    VarIndexer indexer;

    // Visit all trees, iterate on events and statements
    visit( indexer);

    // Get result moved in myVariables
    myVariables = indexer.getVarNames();
  }

  // Access number of variables (vector size) and names
  const vector<string>& varNames() const
  {
    return myVariables;
  }

  // ...
};
```

After a *Product* has been created with its *parseEvents*() method, we call its *indexVariables*() method, which correctly sets indices on all variable nodes in all trees and records the number and names of the variables in the *Product* object. After simulations, we can use the *varNames*() method to get the number, names, and indices of all variables in the script so that we can associate values to names.

Note that the variable indexer, as most visitors, sequentially visits events from today to maturity and all statements included in each event. It carries a state (the map of variables met so far) that changes with the visits and is used for subsequent visits. The state is the reason why visitors are so much more powerful than coding the corresponding actions directly as virtual methods in the node objects.

3.4 PRE-PROCESSORS

The variable indexer is one of the many particular visitors we call *pre-processors*. Pre-processors analyze the trees before simulations take place, anticipate the subsequent work of run-time visitors, and perform actions before simulations start so as to minimize the computational effort of those visitors that work repeatedly during simulations. Pre-processors pre-allocate working memory, index things so they can be directly accessed in memory, and possibly perform some of the forthcoming calculations that don't depend on the simulated data. Pre-processors typically store some results on the nodes of the expression trees, like the variable indexer stores the indices on the variable nodes.

Another example is described in part IV on Fuzzy Logic, where our algorithm identifies the variables affected in *if... then... else... endIf* statements and allocates space for their temporary storage.

Our section III, chapter 13, dedicated to interest rates, discusses a major performance improvement due to a simulated data pre-processor, which identifies what data must be simulated for what date, pre-allocates memory space for that data, and indexes it so it is directly accessed in memory during simulations, like the variables in the script.

Note that variable indexing could have been performed at parsing time, the parser indexing the variables on the fly as it parses them and keeping a map of previously parsed variables as a state. This is the case for most pre-processing we may want to carry out. However, given the large amount of pre-processing we will eventually perform, to do so at parsing time would cause the parsing code to unnecessarily grow in size and complexity and quickly become hard to read and maintain. It is better practice to parse events into raw expression trees first and then have a (potentially large) number of specialized pre-processors, each performing a specific job after parsing and before simulations. All the pre-processors may be called in sequence in a *preProcess*() method of the Product class. At the moment, we have:

```
class Product
{
  // ...

  // Fire all pre-processors
  void preProcess()
  {
    // Index variables
    indexVariables();

    // Call all future pre-processing methods here
  }

  // ...
};
```

We will also see exceptions to this rule in section II on basic improvements, where a lot of small operations can effectively be conducted at parsing time, or even before that, as string manipulations, where the development of dedicated visitors would be overkill.

3.5 CONST VISITORS

The variable indexer modifies the trees it visits when it sets indices on the variable nodes. Our debugger, on the other hand, only traverses trees to pick information and does not modify trees however as it visits them. The debugger is a *const* visitor.

While the distinction is not a major concern in the case of the debugger, which is essentially a development tool, it does matter a lot in the case of the evaluator, described next, which is also a const visitor.

The evaluator is not meant to modify trees on visits. It only reads information stored inside the nodes it traverses and uses this information to perform calculations and store results inside its own data members. The evaluator is a const visitor, and it is important to keep it this way for the sake of thread safety.

When we multi-thread Monte-Carlo simulations over scenarios with scripted transactions, we use different evaluators on each thread, and all these evaluators visit the same trees in parallel. Hence, to avoid race conditions, evaluators must absolutely refrain from modifying the visited trees in any way. And it is advisable to structure the code to guarantee the constness of the visited trees during evaluations.

We therefore declare a *constVisitor* class, along the *Visitor* class, that essentially does the same thing, except it ensures the constness of the visited nodes. There is some code duplication here, so we essentially need to duplicate the definition of the *Visitor* class and the *acceptVisitor*() methods in all concrete *Node* classes. This operation could be templated, but the resulting code would end up quite esoteric, so we settle for an old-fashioned duplication.

We have the declaration of the *constVisitor* class along *Visitor*:

```
class constVisitor
{
protected:
  // Protected constructor so the base class cannot be instantiated
  constVisitor() {}

public:
  virtual ~constVisitor() {}

protected:
  // Default visit just visits arguments
  // so as to ensure that the whole sub-tree is visited
```

```
  virtual void visitArguments( const Node& node)
  {
    for( auto& arg : node.arguments) arg->acceptVisitor( *this);
  }

public:

  // Entry point for visiting a (sub-) tree
  void visit( ExprTree& tree)
  {
    tree->acceptVisitor( *this);
  }

  // All concrete node default visitors,
  // visit arguments unless overridden

  virtual void visitUplus( const NodeUplus& node)
      { visitArguments( node); }
  virtual void visitUminus( const NodeUminus& node)
      { visitArguments( node); }
  virtual void visitAdd( const NodeAdd& node)
      { visitArguments( node); }
  // ...
  virtual void visitVar( const NodeVar& node)
      { visitArguments( node); }
};
```

And we have support for *constVisitors* in the node classes:

```
struct Node
{
  vector<ExprTree> arguments;

  virtual ~Node() {}

  virtual void acceptVisitor( Visitor& visitor) = 0;
  virtual void acceptVisitor( constVisitor& visitor) const = 0;
};

// Unary +
struct NodeUplus : public Node
{
  void acceptVisitor( Visitor& visitor) override;
  void acceptVisitor( constVisitor& visitor) const override;
};

void NodeUplus::acceptVisitor( constVisitor& visitor) const
{
  visitor.visitUplus( *this);
}

// Unary -
struct NodeUminus : public Node
{
  void acceptVisitor( Visitor& visitor) override;
  void acceptVisitor( constVisitor& visitor) const override;
};
```

```
void NodeUminus::acceptVisitor( constVisitor& visitor) const
{
  visitor.visitUminus( *this);
}

// operator+
struct NodeAdd : public Node
{
  void acceptVisitor( Visitor& visitor) override;
  void acceptVisitor( constVisitor& visitor) const override;
};

void NodeAdd::acceptVisitor( constVisitor& visitor) const
{
  visitor.visitAdd( *this);
}

// Etc
```

Visitors that derive from *constVisitor* will not be permitted to modify nodes. Attempts to do so will result in compilation errors. As long as the evaluator derives *constVisitor*, evaluators are safe to visit the same trees concurrently.

3.6 THE EVALUATOR

Evaluation is a particular form of visit. In each Monte-Carlo simulation, this visitor performs the calculations corresponding to the nodes in the trees in the product, iterating over events and statements, and recording the final values of all the variables. This effectively prices products, because a price is the average across scenarios of the final value of the corresponding variable where all events have been evaluated in accordance to some scenario.

The evaluator only interacts with simulations when it visits the spot node.[2] Before we look into this interaction, we inspect the evaluation of simpler, scenario-independent scripts like, for instance, the (utterly useless but hopefully relevant pedagogically)

```
x = log( 100)
y = 3 * x
if y < 50
then y = 50 z = 100
else z = y
endIf
```

[2]For now. We start introducing other scenario-related nodes in chapter 5 on payments in this section and many more in part III.

After this script is executed, the final values of the variables involved are

x = 4.61
y = 50
z = 100

The first statement assigns *log*(100) into *y*. The second statement reads the value of *x*, multiplies it by 3 and assigns the result to *y*, and so on. Hence, the evaluator needs to carry over the values of all variables as it iterates through statements and events. The values of the variables are actually the only state the evaluator needs to carry. Remember that our variable indexer turned names into indices and recorded in the *Product* the number and names of the variables involved. Therefore, we can start our evaluator with:

```
#include <vector>

template <class T>
class Evaluator : public constVisitor
{
  // State
  vector<T>     myVariables;

  // ...

public:

  // Constructor, nVar = number of variables,
  // from Product after parsing and variable indexation
  Evaluator( const size_t nVar) : myVariables( nVar) {}

  // Access to variable values after evaluation
  const vector<T>& varVals() const
  {
    return myVariables;
  }

  // ...
};
```

Note that the evaluator is templated in the number type for its variables. This is because the evaluation participates in the production of values, and the evaluator's variables are active in that production. In order for the evaluation code to be instrumented with AAD, the type of the variables must be templated. This book is not concerned with AAD and focuses on scripting. We could have made the evaluator a standard class using doubles as number types. But then we would need to return to the code and alter it when we instrument it for AAD. Instead, we directly present the templated version of the code, although, as far as this exposé is concerned, readers can mentally replace the template type with doubles. AAD is covered in volume 1 of this series [27].

The evaluation of expressions and conditions is similar to the debugger. For all nodes other than leafs (variables and constants), spot (this one we look into later), and instructions (*if* and *assign* nodes), evaluation consists of the same steps. First, evaluate the arguments (ensuring whole tree traversal), then fetch the results for the arguments, perform the operation according to the concrete type of the current node, and finally store the result so that a parent node may use it. Due to the recursive nature of trees and visits, it is convenient to use stacks to store intermediary results. We have a stack of numbers for expressions and a stack of booleans for conditions. The nature of the stack means that the arguments are picked and processed in the reverse order they are put on the stack. In order to process arguments first to last, we put them on the stack last to first.

```
template <class T> class Evaluator : public constVisitor
{
  // State
  vector<T>     myVariables;

  // Stacks
  quickStack<T>     myDstack;
  quickStack<bool>   myBstack;

  // ...

  // Visit arguments, right to left
  void evalArgs( const Node& node)
  {
for( auto it = node.arguments.rbegin();
      it != node.arguments.rend();
      ++it)
    (*it)->acceptVisitor( *this);
  }

public:
  // ...
};
```

Visits to all nodes except leafs start with a visit to arguments with a call to *evalArgs*(), then pick the values resulting from the evaluation of the arguments from the relevant stack (left to right), evaluate the current node, and push the result onto the stack so that the parent node can pop it. The add node visitor, for instance, acts as follows:

```
void visitAdd( const NodeAdd& node) override
{
  evalArgs( node);
  auto args=pop2();
  myDstack.push( args.first+args.second);
}
```

So as to make the code more compact, we use a helper function *pop2*() that pops two numbers from the stack and returns them in a pair. We expect compilers to inline the function and/or use RVO so it clarifies the code without a run-time penalty.

```
pair<T,T> pop2()
{
  pair<T,T> res;
  res.first = myDstack.top();
  myDstack.pop();
  res.second = myDstack.top();
  myDstack.pop();
  return res;
}
```

Then all the visitors for binary operators take three lines:

```
void visitSubtract( const NodeSubtract& node) override
{
  evalArgs( node);
  auto args=pop2();
  myDstack.push( args.first-args.second);
}
void visitMult( const NodeMult& node) override
{
  evalArgs( node);
  auto args=pop2();
  myDstack.push( args.first*args.second);
}
void visitDiv( const NodeDiv& node) override
{
  evalArgs( node);
  auto args=pop2();
  myDstack.push( args.first/args.second);
}
void visitPow( const NodePow& node) override
{
  evalArgs( node);
  auto args=pop2();
  myDstack.push( pow( args.first, args.second));
}
```

Unaries are even simpler. The unary + leaves its RHS on the stack, the unary – turns it into its opposite.

```
void visitUplus( const NodeUplus& node) override
   { evalArgs( node); }
void visitUminus( const NodeUminus& node) override
   { evalArgs( node); myDstack.top() *= -1; }
```

Functions work similarly:

```
void visitLog( const NodeLog& node) override
{
  evalArgs( node);

  double res = log( myDstack.top());
  myDstack.pop();

  myDstack.push( res);
}
void visitSqrt( const NodeSqrt& node) override
{
  evalArgs( node);

  double res = sqrt( myDstack.top());
  myDstack.pop();

  myDstack.push( res);
}
void visitMax( const NodeMax& node) override
{
  evalArgs( node);

  double M = myDstack.top();
  myDstack.pop();

  for( size_t i=1; i<node.arguments.size(); ++i)
  {
    M = max( M, myDstack.top());
    myDstack.pop();
  }

  myDstack.push( M);
}
void visitMin( const NodeMin& node) override
{
  evalArgs( node);

  double m = myDstack.top();
  myDstack.pop();

  for( size_t i=1; i<node.arguments.size(); ++i)
  {
    m = min( m, myDstack.top());
    myDstack.pop();
  }

  myDstack.push( m);
}
// ...
```

And the condition operators are identical to binaries, except they use the boolean stack instead of the number stack:

```
pair<bool,bool> pop2b()
{
  pair<bool,bool> res;
  res.first = myBstack.top();
  myBstack.pop();
  res.second = myBstack.top();
  myBstack.pop();
  return res;
}

#define EPS 1.0e-15

void visitEqual( const NodeEqual& node) override
{
  evalArgs( node);
  auto args=pop2();
  myBstack.push( fabs( args.first-args.second) < EPS);
}
void visitDifferent( const NodeDifferent& node) override
{
  evalArgs( node);
  auto args=pop2();
  myBstack.push( fabs( args.first-args.second) > EPS);
}
void visitSuperior( const NodeSuperior& node) override
{
  evalArgs( node);
  auto args=pop2();
  myBstack.push( args.first>args.second + EPS);
}
void visitSupEqual( const NodeSupEqual& node) override
{
  evalArgs( node);
  auto args=pop2();
  myBstack.push( args.first>args.second - EPS);
}
void visitInferior( const NodeInferior& node) override
{
  evalArgs( node);
  auto args=pop2();
  myBstack.push( args.first<args.second - EPS);
}
void visitInfEqual( const NodeInfEqual& node) override
{
  evalArgs( node);
  auto args=pop2();
  myBstack.push( args.first<args.second + EPS);
}
void visitAnd( const NodeAnd& node) override
{
  evalArgs( node);
  auto args=pop2b();
  myBstack.push( args.first && args.second);
}
```

```
    void visitOr( const NodeOr& node) override
    {
      evalArgs( node);
      auto args=pop2b();
      myBstack.push( args.first || args.second);
    }

//  ...
```

Constants are trivial; they just push their constant value onto the number stack.

```
    void visitConst (const NodeConst& node) override
    {
      myDstack.push( node.val);
    }
```

But variables are more involved, because they can be either read or written. When a variable node is visited as an RHS, the variable is being read and its value needs to be pushed on the stack. This is identical to constants. But when a variable is visited as an LHS, then the variable is being written into, and it is its address that needs to be fetched. So that the evaluator knows when an LHS variable is being visited, we add a boolean *myLhsVar* data member, which is generally set to false, and is flicked to true while a visit to an assignment node evaluates its LHS variable, and a *myLhsVarAdr* that contains the address of the LHS variable being evaluated:

```
template <class T> class Evaluator : public constVisitor
{
  // State
  vector<T>     myVariables;

  // Stacks
  quickStack<T>   myDstack;
  quickStack<bool>  myBstack;

  // LHS variable being visited?
  bool       myLhsVar;
  T*       myLhsVarAdr;

  // ...
};
```

Then the visitors for variables and assignment[3] are as follows:

```
    void visitAssign( const NodeAssign& node) override
    {
      // Visit the LHS variable
      myLhsVar = true;
```

[3]For now, only assignments write into variables. We will have more statements writing into variables as we extend the language, and we need to remember to flick *myLhsVar* when these are being visited, too.

```
        node.arguments[0]->acceptVisitor( *this);
        myLhsVar = false;

        // Visit the RHS expression
        node.arguments[1]->acceptVisitor( *this);

        // Write result into variable
        *myLhsVarAdr = myDstack.top();
        myDstack.pop();
    }

    void visitVar( const NodeVar& node) override
    {
        // LHS?
        if( myLhsVar) // Write
        {
            // Record address in myLhsVarAdr
            myLhsVarAdr = &myVariables[node.index];
        }
        else    // Read
        {
            // Push value onto the stack
            myDstack.push( myVariables[node.index]);
        }
    }
```

In the assignment visitor, we flick *myLhsVar* to true during the visit to the LHS variable, then flick it back to false. So when *visitVar*() is executed on the LHS variable node, it evaluates it in LHS mode, recording the address of the variable on the evaluator in *myLhsVarAdr*. When the assignment visitor next visits the RHS expression, with *myLhsVar* flicked back to false, when variables are encountered, *visitVar*() is executed in RHS mode, reading variable values, and pushing them onto the stack, like constants. When the evaluator returns from the visit to the RHS the result of the RHS expression is calculated and stored on the top of the number stack. We pop it from the stack and write it into the address of the LHS variable, updating the evaluator's state. Note that the index of the variables, that is their position in the state of the evaluator, sits on the node where it had been previously, set by the variable indexer.

We only have *visitIf*() left to discuss before we move on to the spot node and the matter of the communication with models. When we visit an *if* node, we start by a visit to the condition (argument 0). The visit to the top of the condition tree recursively visits the whole tree beneath it that represents the whole condition. The result of the evaluation (true or false) ends up on the top of the booelan stack. We pick it from there and evaluate the relevant statements by a visit to their top nodes, according to the result of the condition. Implementation follows.

```
    void visitIf( const NodeIf& node) override
{
    // Visit the condition
    node.arguments[0]->acceptVisitor( *this);

    // Pick the result
    const bool isTrue = myBstack.top();
    myBstack.pop();

    // Evaluate the relevant statements
    if( isTrue)
    {
      const auto lastTrue =
              node.firstElse == -1?
              node.arguments.size()-1:
              node.firstElse-1;
      for( auto i=1; i<=lastTrue; ++i)
      {
        node.arguments[i]->acceptVisitor( *this);
      }
    }
    else if( node.firstElse != -1)
    {
      for( auto i=node.firstElse;
          i<=node.arguments.size()-1;
          ++i)
      {
        node.arguments[i]->acceptVisitor( *this);
      }
    }
}
```

We are left with only one visitor method to write in the evaluator, the one for the visit of the *spot* nodes. This is the only place for now where the scripting library accesses the data simulated by the model. For this reason, before we write the visitor method, we need to design the communication channels between the scripting library and the simulation models.

3.7 COMMUNICATING WITH MODELS

Our scripting language has been purposely designed with loose coupling to the simulation models that may use it. So far, there has been no mention of the simulation model and the entire scripting code has been designed independently. Now is the time where we finally build a bridge between the scripting library and models. We want to keep it generic, minimal, and all in one place in order to preserve the loose coupling. In our framework, only the evaluator needs to communicate with models, and the only thing we need from models is an access to the simulated data on event dates. Communication is needed wherever the evaluator accesses the simulated data. So far in our language, this only happens in the visit to the spot node, and the only required simulated data is the spot price on the corresponding event date. This will change when we extend the language, so we declare a data structure to hold all the necessary simulated data for one event date:

```
template <class T>
struct SimulData
{
  T spot;
};
```

and we define a scenario as an array of these, one for each event date:

```
template <class T>
using Scenario = vector<SimulData<T>>;
```

(Note the templated number type—all calculation code linked to the model is templated so it is ready for AAD; readers may ignore that for now and mentally replace template types with doubles.)

All we need from models is a sequence of scenarios, one per simulation. In order to deliver a scenario, the model needs to know the event dates before simulations take place and set itself up accordingly. This information is held in the Product class and accessible from its *eventDates*() accessor.

We conveniently define an interface for all models that communicate with our scripting module in the form of a mix-in class. This means that we kind of define a model API. We declare it with two functionalities:

1. Before simulations, initialization from a sequence of event dates
2. During simulations, access to scenarios

```
template <class T>
struct ScriptModelApi
{
  virtual void initForScripting(const vector<Date>& eventDates) = 0;

  virtual void nextScenario(Scenario<T>& s) = 0;
};
```

The first method is for the model to set itself up in accordance to the event dates for which it will be required to communicate simulated data. The second method fills a pre-allocated scenario (sequence of simulated data by event date) with values from the next simulation.

In order to access the simulated data, the evaluator must hold a reference to the scenario in the current simulation. It also needs to know the index of the event being valued so that it can access the simulated spot for the right event date.

```
template <class T> class Evaluator : public constVisitor
{
  // State
  vector<T>      myVariables;

  // Stacks
  quickStack<T>      myDstack;
  quickStack<bool>     myBstack;

  // LHS variable being visited?
  bool       myLhsVar;
  T*       myLhsVarAdr;

  // Reference to current scenario
  const Scenario<T>*    myScenario;

  // Index of current event
  size_t      myCurEvt;

// ...

public:

  // ...

  // Set reference to current scenario
  void setScenario( const Scenario<T>* scen)
  {
    myScenario = scen;
  }

  // Set index of current event
  void setCurEvt( const size_t curEvt)
  {
    myCurEvt = curEvt;
  }

  // ...
};
```

After that, the spot node visitor is trivial. Just like the const node, it pushes a value onto the number stack, but this value is not a constant, it is the spot value in the set scenario for the set event date index:

```
void visitSpot( const NodeSpot& node) override
{
  myDstack.push( (*myScenario)[myCurEvt].spot);
}
```

To complete our evaluator, we write a public initializer that resets the evaluator in between simulations, in particular, resetting all variable values to 0. (We also provide the copy and move constructors and assignments):

```
// Copy/Move

Evaluator( const Evaluator& rhs) : myVariables( rhs.myVariables) {}
Evaluator& operator=( const Evaluator& rhs)
{
  if( this == &rhs) return *this;
  myVariables = rhs.myVariables;
  return *this;
}

Evaluator( Evaluator&& rhs) : myVariables( move( rhs.myVariables)) {}
Evaluator& operator=( Evaluator&& rhs)
{
  myVariables = move( rhs.myVariables);
  return *this;
}

// (Re-)initialize before evaluation in each scenario
void init()
{
  for( auto& varIt : myVariables) varIt = 0.0;
  // Stacks should be empty, if this is not the case we empty them
  //  without affecting capacity for added performance
  while( !myDstack.empty()) myDstack.pop();
  while( !myBstack.empty()) myBstack.pop();
  myLhsVar = false;
  myLhsVarAdr = nullptr;
}
```

Finally, we write a public method in our *Product* class for the evaluation of a whole product given a scenario, and, for convenience, factory methods for the generation of correctly sized evaluators and empty scenarios:

```
class Product
{
  // ...

  template <class T>
  void evaluate( const Scenario<T>& scen, Evaluator<T>& eval) const
  {
    // Set scenario
    eval.setScenario( &scen);

    // Initialize all variables
    eval.init();

    // Loop over events
    for( auto i=0; i<myEvents.size(); ++i)
    {
      // Set current event
      eval.setCurEvt( i);
```

```
    // Loop over statements in event
    for( auto& statIt : myEvents[i])
    {
      // Visit statement
      eval.visit( statIt);
    }
  }
}

// Evaluator factory
template <class T>
    unique_ptr<Evaluator<T>> buildEvaluator()
{
  // Move
  return unique_ptr<Evaluator<T>>
      ( new Evaluator<T>( myVariables.size()));
}

// Scenario factory
template <class T>
    unique_ptr<Scenario<T>> buildScenario()
{
  // Move
  return unique_ptr<Scenario<T>>
      ( new Scenario<T>( myEventDates.size()));
}

// ...
};
```

Our basic scripting language is now complete. The *Product* class provides a convenient entry point for working with it.

Its *parseEvents*() method turns event strings indexed by event dates into expression trees and stores the resulting trees internally. Its *indexVariables*() method computes and stores a vector of all variables involved and sets their index in the variable nodes. It may be called as a part of the pre-Process method that calls all necessary pre-processors (only *indexVariables*() for now) in a sequence.

Once the *Product*() is built with a call to *parseEvents*() followed by a call to *indexVariables*() or *preProcess*(), we read the vector of event dates with the accessor *eventDates*(). The model needs this information so it knows for what dates simulated data must be provided during simulations. Our model API takes it as a parameter for the initialization of the model. The vector of variable names is accessed with the *Product*'s accessor *varNames*(). We need these to match results to variable names when pricing/risk is complete. The *buildEvaluator*() factory method provides a correctly initialized *Evaluator*. The *buildScenario*() factory provides a correctly sized empty scenario (vector of *simulData* structs) for the model to fill. Our model API takes the empty scenario as a byRef parameter to be filled by the model simulation by simulation.

For each simulation, we need the model to fill the *Scenario* with simulated data for all event dates. Then we call the *Product*'s *evaluate*() method, passing the generated scenario and the evaluator. The product's *evaluate*() method runs the script and leaves the variables in the evaluator at their final value, in accordance with the scenario. These values are accessed with the evaluator's *varVals*() accessor, which returns a vector of variable values with the same indices as the variable names in *varNames*().

The library is ready to be put together with a model. In the next chapter, we write a simplistic Black-Scholes simulator and show how to link it to our scripting language.

CHAPTER 4

Putting Scripting Together with a Model

4.1 A SIMPLISTIC BLACK-SCHOLES MONTE-CARLO SIMULATOR

Although it is not our purpose here to discuss the design of simulation models, we need a simplistic one to test run our scripting library. For that reason, we develop a simplistic simulation library, which is self-contained and features its own API that is not related in any way to our scripting library. Then we show how to connect the two.

The three main pieces of a simulation library are the random number generator, the model, and the simulation engine.

4.1.1 Random number generators

The random number generator produces the random numbers used for the generation of Monte-Carlo paths:

```
#include <random>
#include <vector>
#include <memory>
using namespace std;

struct randomgen_error : public runtime_error
{
  randomgen_error(const char msg[]) : runtime_error(msg) {}
};

class RandomGen
{
public:

    // Initialise for a given dimension
    virtual void init( const size_t dim) = 0;

    // Generate the next random point
    virtual void genNextNormVec() = 0;
```

```
    // Access Gaussian vector byRef
    virtual const vector<double>& getNorm() const = 0;

    // Clone
    virtual unique_ptr<RandomGen> clone() const = 0;

    // Skip ahead (for parallel Monte-Carlo)
    virtual void skipAhead(const long skip)
    {
        throw randomgen_error
( "Concrete random generator cannot be used for parallel simulations");
    }
};
```

It is initialized with a dimension *dim*, that is, the number of random numbers needed to generate one path. Then, successive calls to *genNextNormVec*() generate the next set of *dim* independent standard Gaussian random numbers, which are accessed by *getNorm*(). It is also useful to implement the *virtual copy constructor* pattern with a cloning function. In addition, random generators used in parallel simulations (not discussed here) must implement a *skip ahead*, that is, the ability to skip a number of random vectors without generating them. Only random generators that are meant to be used in parallel simulations must override this method.

In the interest of self-containment, we provide here a basic generator based on C++11's native facilities, although we strongly advise against its use in production. We recommend reading Numerical Recipes [25], Jackel's Monte-Carlo textbook [19], or Savine's Parallel Simulations lecture notes [27] for discussions of efficient random number generators.

```
class BasicRanGen : public RandomGen
{
  default_random_engine myEngine;
  normal_distribution<> myDist;
  size_t   myDim;

  vector<double>  myNormVec;

public:

  BasicRanGen( const unsigned seed = 0)
  {
    myEngine = seed > 0?
        default_random_engine( seed):
        default_random_engine();
    myDist = normal_distribution<>();
  }
```

```
    void init(const size_t dim) override
    {
        myDim = dim;
        myNormVec.resize(dim);
    }

  void genNextNormVec() override
  {
    for( size_t i=0; i<myDim; ++i)
    {
      myNormVec[i] = myDist( myEngine);
    }
  }

    const vector<double>& getNorm() const override
  {
    return myNormVec;
  }

    // Clone
    unique_ptr<RandomGen> clone() const override
    {
        return unique_ptr<RandomGen>
            (new BasicRanGen(*this));
    }
};
```

4.1.2 Simulation models

The second piece is the simulation model, which has the sole responsibility to turn random numbers into a Monte-Carlo path:

```
template <class T>
struct Model
{
      // Clone
   virtual unique_ptr<Model> clone() const = 0;

   // Initialize simulation dates
   virtual void initSimDates(const vector<Date>& simDates) = 0;

   // Number of Gaussian numbers required for one path
   virtual size_t dim() const = 0;

   // Apply the model SDE
   virtual void applySDE(
      // Gaussian numbers, dimension dim()
      const vector<double>&  G,
      // Populate spots for each event date
      vector<T>&  spots)
         const = 0;
};
```

This design fits only single-asset models such as equity, forex, or commodity. It is not sufficient to work with interest rates, but it does accommodate any type of volatility model: constant, time-dependent, local, stochastic, with and without jumps, and so on.

The model must be fed with event dates, the dates for which it is required to produce simulated market variables (in our simple case, spots). After that initialization, the model must provide its dimension (the number of random numbers required to generate one path). Note that the dimension is not necessarily the number of event dates. In a stochastic volatility model, for instance, the dimension is at least twice the number of event dates, since random increments are needed for the path of the volatility, too. In addition, the simulation timeline of the model may include more dates than the event dates for those models that need to discretize SDEs over short time periods. Finally, the model provides a method *applySDE()* that takes a vector of random numbers of dimension dim and uses them to generate a path for the spot of dimension the number of event dates. The number type in the model is templated so the model is suitable to AAD. The random numbers, however, are not templated, since they are inactive (see for instance Flyger, Huge, and Savine's AAD presentation [11]).

For instance, the following instantiates a simple Black-Scholes model:

```
template <class T>
class SimpleBlackScholes : public Model<T>
{
   Date myToday;

   T mySpot;
   T myVol;
   T myDrift;

   bool  myTime0; // If today is among simul dates
   vector<double> myTimes;
   vector<double> myDt;
   vector<double> mySqrtDt;

public:

   // Construct with T0, S0, vol and rate
   SimpleBlackScholes( const Date& today,
                  const double spot,
                  const double vol)
   : myToday( today), mySpot( spot), myVol( vol),
      myDrift( 0.5*vol*vol)
   {}

   // clone
   virtual unique_ptr<Model> clone() const override
   {
  return unique_ptr<Model>(new SimpleBlackScholes(*this));
   }
```

```
    //  Parameter accessors, read only
    const T& spot() { return mySpot; }
    const T& vol() { return myVol; }

    // Initialize simulation dates
    void initSimDates(const vector<Date>& simDates) override
    {
      myTime0 = simDates[0] == myToday;

      // Fill array of times
      for( auto dateIt = simDates.begin();
          dateIt != simDates.end();
          ++dateIt)
      {
        myTimes.push_back( double( *dateIt - myToday) / 365);
      }
      myDt.resize( myTimes.size());
      myDt[0] = myTimes[0];
      for( size_t i=1; i<myTimes.size(); ++i)
      {
        myDt[i] = myTimes[i] - myTimes[i-1];
      }
      mySqrtDt.resize( myTimes.size());
      for(size_t i=0; i<myTimes.size(); ++i)
      {
        mySqrtDt[i] = sqrt( myDt[i]);
      }
    }

    size_t dim() const override { return myTimes.size() - myTime0; }

    // Simulate one path
    void applySDE(
        // Gaussian numbers, dimension dim()
        const vector<double>&  G,
        // Populate spots for each event date
        vector<T>&  spots)
        const override
    {
        //  Apply the SDE
        size_t step = 0;

      // First step
      spots[0] = myTime0?
      mySpot:
      mySpot*exp(-myDrift*myDt[0]+myVol*mySqrtDt[0]*G[step++]);

      // All steps
      for(size_t i=1; i<myTimes.size(); ++i)
      {
        spots[i] = spots[i-1]
            *exp(-myDrift*myDt[i]+myVol*mySqrtDt[i]*G[step++]);
      }
    }
};
```

Everything in this code should be self-explanatory. Note the technicality of a special treatment in case the model's today date is present among the event dates.[1]

4.1.3 Simulation engines

Finally, a Monte-Carlo simulator brings together a random generator and a model:

```
template <class T>
class MonteCarloSimulator
{
    RandomGen&   myRandomGen;
    Model<T>&   myModel;

public:

    MonteCarloSimulator( Model<T>& model, RandomGen& ranGen)
: myRandomGen( ranGen), myModel( model) {}

    void init( const vector<Date>& simDates)
    {
        myModel.initSimDates( simDates);
        myRandomGen.init( myModel.dim());
    }

    void simulateOnePath( vector<T>& spots)
    {
        myRandomGen.genNextNormVec();
        myModel.applySDE(myRandomGen.getNorm(), spots);
    }
};
```

Once again, the implementation should be self-explanatory.

4.2 CONNECTING THE MODEL TO THE SCRIPTING FRAMEWORK

In order to use this model, or any other model, either purposely developed, or existing, with our scripting library, we must instantiate it with the API we defined previously. This API is a mix-in class, the equivalent to a Java interface in C++. This means that a model that is able to communicate with our module must inherit from both itself and the scripting API and implement the two virtual functionalities declared in the API.

[1]Our implementation does not deal with past dates, something that would be necessary in a production system.

```
template <class T>
class ScriptSimulator :
   public MonteCarloSimulator<T>, public ScriptModelApi<T>
{

   vector<T> myTempSpots;

public:

   ScriptSimulator( Model<T>& model, RandomGen& ranGen)
      : MonteCarloSimulator( model, ranGen) {}

   void initForScripting( const vector<Date>& eventDates) override
   {
      MonteCarloSimulator::init( eventDates);
      myTempSpots.resize( eventDates.size());
   }

   void nextScenario( Scenario<T>& s) override
   {
      MonteCarloSimulator::
         simulateOnePath( myTempSpots);

      // Note the inefficiency
      for(size_t i=0; i<s.size(); ++i) s[i].spot = myTempSpots[i];
   }
};
```

That's all. We just need a driver to put it all together, for instance, this one, below.

```
void simpleBsScriptVal(
  const Date&    today,
  const double   spot,
  const double   vol,
  const map<Date,string>& events,
  const unsigned   numSim,
  // Results
  vector<string>&   varNames,
  vector<double>&   varVals)
{
  if( events.begin()->first < today)
   throw runtime_error("Events in the past are disallowed");

  // Initialize product
  Product prd;
  prd.parseEvents( events.begin(), events.end());
  prd.indexVariables();

  // Build evaluator and scenarios
  unique_ptr<Scenario<double>> scen = prd.buildScenario<double>();
  unique_ptr<Evaluator<double>> eval = prd.buildEvaluator<double>();

  // Initialize model
  BasicRanGen random;
   SimpleBlackScholes<double> model(today, spot, vol);
```

```
    ScriptSimulator<double> simulator( model, random);
      simulator.initForScripting( prd.eventDates());

  // Initialize results
  varNames = prd.varNames();
  varVals.resize( varNames.size(), 0.0);

  // Loop over simulations
  for( size_t i=0; i<numSim; ++i)
  {
    // Generate next scenario into scen
    simulator.nextScenario( *scen);
    // Evaluate product
    prd.evaluate( *scen, *eval);
    // Update results
    for( size_t v=0; v<varVals.size(); ++v)
    {
      varVals[v] += eval.varVals()[v] / numSim;
    }
  }
}
```

Note that we disallowed past events in this simple example. In real production systems based on scripting, however, cash-flows need to be dealt with throughout their life cycle. For example, variables reflecting path dependency need to be updated from historical data. In production systems, models must fill past data from a database and future data from simulations. How this is implemented is system dependent and out of scope here. This is not a difficult development and one that does not interfere with the algorithms exposed here.

We now have a simple, fully functional, framework for the valuation of scripts in a simulation model. The software is most conveniently used as an Excel add-in. See for instance Dalton's publication [8] for a complete overview of exporting C++ to Excel. Our online repository also has a short tutorial and code for exporting C++ functions to Excel. In the repository, the function is exported to Excel. Here in the book, we build it into a console application for completion.

```
int evalTester()
{
  // Inputs
  string str;
  Date today;
  double spot, vol;
  unsigned numSim;
  cout << "Enter Todays date: ";
  getline( cin, str);
  today = stoi( str);
  cout << "Enter spot: ";
  getline( cin, str);
```

```
spot = stod( str);
cout << "Enter vol: ";
getline( cin, str);
vol = stod( str);
cout << "Enter num sim: ";
getline( cin, str);
numSim = stoi( str);

// Build the event map with user inputs

Date date;
string evtStr;
map<Date,string> events;

cout << "Enter events, -1 when done" << endl;

cout << "Enter date: ";
getline( cin, str);
date = stoi( str);

if( date != -1)
{
  cout << "Event event for date " << date << " >> ";
  getline( cin, evtStr);
}

while( date != -1)
{
  events[date] = evtStr;

  cout << "Enter date: ";
  getline( cin, str);
  date = stoi( str);

  if( date != -1)
  {
    cout << "Event event for date " << date << " >> ";
    getline( cin, evtStr);
  }
}

// Main call
vector<string> varNames;
vector<double> varVals;
try
{
  simpleBsScriptVal(
    today,
    spot,
    vol,
    events,
    numSim,
    varNames,
    varVals);
}
```

```
    catch (const runtime_error& rte)
    {
        cout << rte.what();
        return 1;
    }

    for( unsigned i=0; i<varNames.size(); ++i)
    {
        cout << varNames[i] << " = " << varVals[i] << endl;
    }

    return 0;
}
```

Importantly, we used a simplistic Black-Scholes model here, but you may use a local volatility model instead or your favorite stochastic volatility model, or any other model that already exists in your library or one that you develop purposely. All you need to do is instantiate your model against the API and everything falls in place. No change whatsoever is necessary in the scripting code.

At present, however, we are limited to one underlying and this one underlying cannot be an interest rate. We discuss rates and multiple underlyings in part III.

In addition, our code presently accepts models only with constant zero interest rates. We address this limitation now, with the development of an important extension. And we take advantage of this development to illustrate the necessary steps for a core extension to our language.[2]

A last word on visitors: *there is a visitor for everything*. We met a visitor with only a few lines of code that lists and indexes all variables involved in a script. In the context of xVA, we developed visitors that are able to compress cash-flows from multiple transactions and *decorate* a script to compute the payoff of an xVA together with the existing scripted cash-flows in a netting set. Visitors can analyze products in many ways that may be useful for modeling and also for back-office processing, back or forward testing, or the management of a product throughout its life cycle. For example, a visitor can analyze dependencies between variables in a script, list all upcoming payments and coupons, or process a script and update it on corporate events.

[2]A trivial extension would be, for instance, to add some functions. What we discuss next are core extensions with new keywords representing new concepts.

CHAPTER 5

Core Extensions and the "Pays" Keyword

Before we close this section, we implement support for basic interest rates and cash-flow discounting. Support for payments that depend on interest rate fixings is discussed later; for now we only consider the discounting of payments. We however consider discounting in a very general way so as to include stochastic discounting, coming for example from a Black-Scholes or Dupire model with stochastic interest rates. In addition to being an important extension in itself, this paragraph may be used as a reference on the necessary steps for the implementation of a core extension to the scripting language.

5.1 IN THEORY

In the absence of interest rates, a product that pays n event-driven cash-flows $C_1, \dots, C_n$ on dates $T_1, \dots, T_n$ has value $V = E^{RN}[sum_i\ C_i]$, where the expectation is computed under the risk-neutral probability measure. Asset prices are martingales under this measure unless they pay dividends or other benefits.

When models support interest rates, they are often written under the so-called risk-neutral probability measure or, equivalently, under the bank account *numeraire*. In this case, we simulate the bank account $B(t) = \exp\left(\int_0^t r_s ds\right)$ together with the spot, where r is the instantaneous rate, and value products with $V = E^{RN}\left[\sum_{i=1}^{n} \frac{C_i}{B(T_i)}\right]$. Then, the risk-neutral drift of assets that don't pay dividends is the rate r. Note that it is the responsibility of the model to correctly implement drifts. This is not something scripting needs to worry about.

Model developers may opt for a diffusion under a different numeraire N. In this case, we have $V = N_0 E^N\left[\sum_{i=1}^{n} \frac{C_i}{N_i}\right]$, or, setting $N_0 = 1$ without loss of generality, $V = E^N\left[\sum_{i=1}^{n} \frac{C_i}{N_i}\right]$. In this general case, expectations are taken under the *martingale measure associated with the numeraire* N and under this measure, it is the ratio of non-dividend-paying asset prices to the numeraire that are martingales. It is the model's responsibility to correctly implement the dynamics of the asset prices and the numeraire. But it is up to the scripting language to correctly discount payments.

In practice, to support discounting in a general way, whatever the numeraire in the simulation, the model must provide the numeraire, in addition to the asset price, for all event dates in every simulation, as an additional simulated data. Then, on the payment of a coupon, the scripting module must discount the payment by the provided numeraire. For instance, for a 1Y100 European call, we could write the script (today is 1 June 2020):

```
01Jun2021    opt = max( spot() - 100, 0) / numeraire()
```

where *numeraire*(), like *spot*(), is a keyword for the access to simulated data. This properly implements discounting in the language and the communication with models in a general way. However, the syntax is rather clumsy and somewhat unnatural. And what to say of a cliquet, scripted here:

01Jun2020	vRef = spot()
01Jun2021	vPerf = Max(spot() - vRef, 0) / vRef prd = vPerf / numeraire() vRef = spot()
01Jun2022	vPerf = Max(spot() - vRef, 0) / vRef prd = prd + vPerf / numeraire() vRef = spot()
01Jun2023	vPerf = Max(spot() - vRef, 0) / vRef prd = prd + vPerf / numeraire()

This is correct, but very clumsy. We suggest a more elegant solution based on a new keyword *pays*.

```
prd pays expression
```

as a shortcut for:

```
prd = prd + expression / numeraire()
```

Since all variables are initialized to 0, this is a general syntax that works with single payments, too. This new keyword *pays* encapsulates the discounting logic so that users never need to worry about it. It also elegantly resolves the scripting of products with multiple payments. Assignment is no longer used for payments, only for ancillary variables. Payments and coupons always use *pays* instead.

This syntax offers benefits beyond user-friendliness. A visitor stopping on the pays nodes can figure what variables are products and what are helpers. Crucially for xVA, it can pick all payments and the corresponding payment dates. This would allow visitors to compress cash-flows, and analyze associated credit risks. The credit risk of a coupon does not stop on its fixing; it runs all the way to payment. For instance, in

FixingDate	vCpn = ...
PayDate	prd pays vCpn

there would be no way to determine the exact maturity of the credit risk without the *pays* keyword. With this keyword, a dedicated visitor can figure the actual payment dates and let a model analyze credit risk accordingly.

The ability of visitors to traverse scripts and pick payments may also help with back-office processing and the management of the life cycle of the product. Such particular visitors for payments would need to visit only the *pays* nodes.

5.2 IN CODE

The addition of the keyword *pays* is a fundamental modification of our language because it adds a third type of instruction, hence, a third type of statement. Hopefully, however, the designs we put in place may accommodate such developments without major difficulty. Let us discuss the steps involved in this development.

First, we need a model that supports discount rates and communicates numeraires. We need the numeraire to be part of our simulated data:

```
template <class T>
struct SimulData
{
  T  spot;
  T  numeraire;
};
```

Models must provide numeraires (or 1 if they cannot, this is all part of their specific API wrapper). We support discounting in a fully general way, so that for instance a model *a la Dupire* with stochastic rates would be supported. Here we extend our simplistic Black-Scholes simulator to support a constant rate.

```
template <class T> struct Model
{
   // ...

   virtual void applySDE(
      // Gaussian numbers, dimension dim()
      const vector<double>&  G,
      // Populate spots for each event date
      vector<T>&  spots,
      // Populate numeraire for each event date
      vector<T>&  numeraires)
         const = 0;
};

template <class T> class SimpleBlackScholes : public Model<T>
{
   Date myToday;

   T mySpot;
   T myRate;
   T myVol;
   T myDrift;

// ...

private:

   // Calculate all deterministic discount factors
   void calcDf( vector<T>& dfs) const
   {
      for (size_t i = 0; i<myTimes.size(); ++i)
         // Despite the name of the variable, this is not the discount factor
         // but its inverse, hence, the numeraire
         dfs[i] = exp(myRate * myTimes[i]);
   }

public:

   // Construct with T0, S0, vol and rate
   SimpleBlackScholes( const Date& today,
                  const double spot,
                  const double vol,
                  const double rate)
   : myToday( today), mySpot( spot), myVol( vol), myRate( rate),
   myDrift( -rate+0.5*vol*vol)
   {}

   const T& spot() { return mySpot; }
   const T& rate() { return myRate; }
   const T& vol() { return myVol; }
```

```
// ...

   void applySDE(
      // Gaussian numbers, dimension dim()
      const vector<double>&  G,
      // Populate spots for each event date
      vector<T>&  spots,
      // Populate numeraire for each event date
      vector<T>&  numeraires)
         const override
   {
      // Compute discount factors
      calcDf( numeraires);
      // Note the ineffiency:
      // in this case, numeraires could be computed only once

      // Then apply the SDE
      // ... Nothing to change here, the rate was made part of myDrift
};

template <class T>
class MonteCarloSimulator
{
// ...

public:

// ...

   void simulateOnePath( vector<T>& spots, vector<T>& numeraires)
   {
      myRandomGen.genNextNormVec();
      myModel.applySDE(myRandomGen.getNorm(), spots, numeraires);
   }
};

template <class T>
class ScriptSimulator :
   public MonteCarloSimulator<T>, public ScriptModelApi<T>
{

   vector<T> myTempSpots;
   vector<T> myTempNumeraires;

public:

   ScriptSimulator( Model<T>& model, RandomGen& ranGen) :
      MonteCarloSimulator( model, ranGen) {}

   void initForScripting( const vector<Date>& eventDates) override
{
      MonteCarloSimulator::init( eventDates);
      myTempSpots.resize( eventDates.size());
      myTempNumeraires.resize(eventDates.size());
   }
```

```
    void nextScenario( Scenario<T>& s) override
{
        MonteCarloSimulator::
            simulateOnePath( myTempSpots, myTempNumeraires);

        // Note the inefficiency
        for(size_t i=0; i<s.size(); ++i)
    {
        s[i].spot = myTempSpots[i];
      s[i].numeraire = myTempNumeraires[i];
    }
  }
};

inline void simpleBsScriptVal(
  const Date&    today,
  const double    spot,
  const double    vol,
  const double    rate,
  const map<Date,string>& events,
  const unsigned    numSim,
  // Results
  vector<string>&    varNames,
  vector<double>&    varVals)
{
// ...

SimpleBlackScholes<double> model(today, spot, vol, rate);

// ...
}
```

That resolves the matter of communication with the model. As for the developments in the scripting language itself, we need to:

1. Create the pay node type and add support for it in the base visitor and *constVisitor* classes.
2. Add support in the parser, at instruction level.
3. Add support in the evaluator and other relevant visitors.[1]

We will see that the designs we put in place make these developments rather painless. First, create the *pays* node type, like the other concrete node types:

```
struct NodePays : public Node
{
  void acceptVisitor( Visitor& visitor) override;
  void acceptVisitor( constVisitor& visitor) const override;
};
```

[1]There is nothing to modify in the variable indexer, since this one visits only variables. We may want to extend the debugger.

```
void NodePays::acceptVisitor( Visitor& visitor)
{
  visitor.visitPays( *this);
}

void NodePays::acceptVisitor( constVisitor& visitor) const
{
  visitor.visitPays( *this);
}
```

With support in the base *Visitor* and *constVisitor* classes:

```
class Visitor
{
// ...
  virtual void visitPays( NodePays* node)
     { visitArguments( node); }
// ...
};

class constVisitor
{
// ...
  virtual void visitPays( const NodePays& node)
     { visitArguments( node); }
// ...
};
```

Second, in the parser, add support for the *pays* keyword in the statement parser:

```
static Statement parseStatement( TokIt& cur, const TokIt end)
{
  // Check for instructions of type 1, so far only 'if'
  if( *cur == "IF") return parseIf( cur, end);

  // Parse cur as a variable
  auto lhs = parseVar( cur);

  // Check for end
  if( cur == end) throw script_error( "Unexpected end of statement");

  // Check for instructions of type 2, so far only assignment
  if( *cur == "=") return parseAssign( cur, end, lhs);
  else if( *cur == "PAYS") return parsePays( cur, end, lhs);

  // No instruction, error
  throw script_error( "Statement without an instruction");
  return Statement();
}
```

And implement the parser for *pays*:

```
static ExprTree parsePays( TokIt& cur, const TokIt end, ExprTree& lhs)
{
  // Advance to token immediately following "pays"
  ++cur;

  // Check for end
  if( cur == end) throw script_error( "Unexpected end of statement");

  // Parse the RHS
  auto rhs = parseExpr( cur, end);

  // Build and return the top node
  return buildBinary<NodePays>( lhs, rhs);
}
```

Finally, program the payment visitor in the evaluator. Remember to flick *myLhsVar* when visiting the product making the payment:

```
void visitPays( const NodePays& node) override
{
  // Visit the LHS variable
  myLhsVar = true;
  node.arguments[0]->acceptVisitor( *this);
  myLhsVar = false;

  // Visit the RHS expression
  node.arguments[1]->acceptVisitor( *this);

  // Write result into variable
  *myLhsVarAdr += myDstack.top() / (*myScenario)[myCurEvt].numeraire;
  myDstack.pop();
}
```

Note the close similarity to the assignment. *pays* is basically an assignment, with the difference that it adds to variables rather than overwriting them and normalizes the RHS expression by the numeraire.

That is all. Our language now supports discounting and accepts models with (possibly stochastic) rates. It elegantly supports multiple payments with a natural syntax. We have visitable pays nodes for all payments so that payments may be picked and analyzed by visitors. The whole extension on the scripting side took just over 10 lines of code.

This concludes our basic scripting language. As promised, we developed a fully functional self-contained module that can be used with a variety of models. However, our language cannot deal with interest rates other than for discounting. It cannot deal with multiple underlying assets, like with basket options or hybrid structured products. It cannot handle payments in multiple currencies. It has no support for advanced features like continuous barriers or early exercise with LSM. And it has no support for dates, forcing a cliquet to script as (today is 1 June 2020):

01Jun2020	vRef = spot()
01Jun2021	vPerf = Max(spot() - vRef, 0) / vRef prd pays vPerf vRef = spot()
01Jun2022	vPerf = Max(spot() - vRef, 0) / vRef prd pays vPerf vRef = spot()
01Jun2023	vPerf = Max(spot() - vRef, 0) / vRef prd = pays vPerf

Whereas it really should take the compact script:

01Jun2020	vRef = spot()
start: 01Jun2020 end: 01Jun2023 freq: 1y fixing: end	vPerf = Max(spot() - vRef, 0) / vRef prd pays vPerf vRef = spot()

These are features that bring our scripting play tool to production standards. Luckily, our code was written with these extensions in mind, and it sits on solid foundations. To extend it into a fully professional module for production risk management does not take a full rewriting, but hopefully painless extensions, like what we just did with *pays*. We discuss these extensions in the following sections, although we no longer provide the full source code. Readers with a good understanding of this section should be in a position to implement these extensions discussed without major difficulty.

The complete source code is available in our repository.

PART II

Basic Improvements

Introduction

In this section, we discuss some cosmetic improvements to our scripting language that in some cases can make major differences for the users. For now, we do not deal with support for multiple underlying assets or interest rates. We stick to a single underlying of equity, forex, or commodity type and describe additional features in the scripting language that make it more relevant and practical in a production context.

The source code for the improvements discussed here is not provided, either in the text or in the online repository. Their implementation is, however, discussed in depth. For the features discussed here, the implementation mostly consists of high-level developments that modify the script before it is parsed and only occasionally need minor changes in the parsing code, definition of expression tree nodes, and visitors. We show that only light developments are required to turn our basic scripting library into a professional tool suitable for production. Support for interest rates and multiple underlying assets is more involved and considered in part III.

CHAPTER 6

Past Evaluator

One development needed for production is the support for past events. Once a product like a barrier option is booked, the script is meant to remain unchanged throughout its life cycle. When its value and risk sensitivities are recalculated at a later date, some events, like barrier monitoring, occur in the past. Simulations always start on the valuation date. To evaluate past events, we need another visitor, a dedicated pre-processor that evaluates past events before simulations take place. This pre-processor is actually an evaluator, and works exactly like the ordinary evaluator, except it picks market data on a historical database instead of simulated scenarios. In other words, this is a "past evaluator" that derives from the evaluator and overrides the visits to nodes that pick market data (so far, only the spot node). The final state of the past evaluator after all past events are evaluated is the initial state to which the actual evaluator is reset on each simulation. The state of an evaluator consists of the value for all variables. Hence, the starting values for all the variables in the simulations are their final values from the evaluation of past events by the derived evaluator. For instance, consider a barrier script that uses *vAlive*, a variable that starts at 1 at the trade date and changes to 0 when the barrier is hit. In evaluating past events, the past event evaluator produces a final value for *vAlive* at the valuation date of 1 if the barrier was not hit, and 0 if it was hit. Let's call *vAlive*0 the value of *vAlive* after past event evaluation. During simulations, the value of *vAlive* is reset to *vAlive*0 before each simulation starts. In addition, the cash-flows paid in the past must be disregarded. So the pays node visitor must also be derived in the past evaluator to flag the payment and instantiate it being sent to the client, rather than adding it to the variable in question. The variables in the main evaluator are no longer reset to 0 before each simulation, but to the output of the past evaluator.

The practical implementation of the past evaluator visitor is somewhat dependent on the system and environment and is not further discussed. We do, however, remark that the past evaluator in one snap eliminates the need for any product or underlying specific code for fixing and payments to clients.

One can construct other types of past evaluator visitors, for example, for monitoring barriers and exercises.

CHAPTER 7

Macros

A first cosmetic improvement is the support for C/C++ like macros. Say we are pricing a 3y cliquet with strike at 5% annual performance (today is 1 June 2020):

01Jun2020	vRef = spot()
01Jun2021	vPerf = (spot() - vRef) / vRef prd pays max(vPerf - 0.05, 0) vRef = spot()
01Jun2022	vPerf = (spot() - vRef) / vRef prd pays max(vPerf - 0.05, 0) vRef = spot()
01Jun2023	vPerf = (spot() - vRef) / vRef prd = pays max(vPerf - 0.05, 0)

If we want to change the strike to 2.5%, we need to replace 0.05 by 0.025 in three different places in the script. This is inconvenient and error prone. It is much neater to define a macro:

```
STRIKE   0.05
```

Then the script becomes:

01Jun2020	vRef = spot()
01Jun2021	vPerf = (spot() - vRef) / vRef prd pays max(vPerf - STRIKE, 0) vRef = spot()

01Jun2022	vPerf = (spot() - vRef) / vRef prd pays max(vPerf - STRIKE, 0) vRef = spot()
01Jun2023	vPerf = (spot() - vRef) / vRef prd = pays max(vPerf - STRIKE, 0)

This is a very simple improvement that effects the clarity and maintainability of the scripts. The parameters of a product are held in one place, where macros are defined under an explicit name that refers to what the parameter stands for: STRIKE, BARRIER, THRESHOLD, and so on. The script uses those names rather than magic numbers, making it tidier and more understandable. We can change parameters without any modification to the core of the script. That means that we can build *script templates* for families of transactions.

The simplest implementation consists of a replacement of all macros by their definitions, before parsing even takes place. This means that the strings that initially represent events are modified, with macros replaced by their definitions, before these strings are sent to our parser. This is similar to C/C++ macros, which are replaced by their definitions before the code is compiled. This means that there is no modification necessary to our scripting code. A call to a new function is made before our library is invoked the inputs to the parser.

This function performs work similar to *ctrl* + *h* in Microsoft Office: find macros in the script and replace them with their definitions. Such work is the *raison d'etre* of the C++11 *regex* library, which provides a regex_replace algorithm that does just that. In addition, the regex library also has all necessary functionality to implement macros with arguments, again, similarly to C/C++ macros. This way, we can write a script such as this:

STRIKE	120
CALL(S, K)	max(0, S-K)
01Jun2021	opt pays CALL(spot(), STRIKE)

But after regex has done its work, our parser receives a single event string "opt pays max(0, spot() – 120)." We refer to numerous textbooks and online tutorials available for the C++11's regex library.

However, if we wish to differentiate the product value with respect to any of the macro-defined constants, for example, to strike or notional, the replacement approach is not sufficient and a dedicated macro or definition node has to be implemented.

CHAPTER 8

Schedules of Cash-Flows

Support for schedules is another light development that makes a very significant difference for the users. Without schedules, we script a monthly monitored 1y barrier option as follows:

Barrier script 1

01Jun2020	vAlive = 1
01Jul2020	if spot() > BARRIER then vAlive = 0 endIf
01Aug2020	if spot() > BARRIER then vAlive = 0 endIf
01Sep2020	if spot() > BARRIER then vAlive = 0 endIf
01Oct2020	if spot() > BARRIER then vAlive = 0 endIf
01Nov2020	if spot() > BARRIER then vAlive = 0 endIf
01Dec2020	if spot() > BARRIER then vAlive = 0 endIf
01Jan2021	if spot() > BARRIER then vAlive = 0 endIf
01Feb2021	if spot() > BARRIER then vAlive = 0 endIf
01Mar2021	if spot() > BARRIER then vAlive = 0 endIf
01Apr2021	if spot() > BARRIER then vAlive = 0 endIf
01May2021	if spot() > BARRIER then vAlive = 0 endIf
01Jun2021	if spot() > BARRIER then vAlive = 0 endIf opt pays vAlive * CALL(spot(), STRIKE)

The problem is evident. And what if monitoring was weekly? Or daily? The solution is evident, too: we put repeated cash-flows *on a schedule*:

Barrier script 2

01Jun2020	vAlive = 1
start: 01Jun2020 end: 01Jun2021 freq: 1m fixing: end	if spot() > BARRIER then vAlive = 0 endIf
01Jun2021	opt pays vAlive * CALL(spot(), STRIKE)

A schedule consists of a start date, an end date, and a frequency. Another parameter to the schedule is the fixing. In this case, we set the fixing in the end. This means that the event date (or fixing date) for each period is at the end of the period. For instance, the first period on the schedule starts on 01Jun2020 and ends on 01Jul2020. We put the event date for that first period on July, at the end of the period. The first event date is not on the start date (June), but on the second date (July) on the schedule. The last event date is on the end date of the schedule: 01Jun2021.

An alternative is to fix at the start of the period. In this case, event dates are put on the start dates of the periods. For the first period running from 01Jun2020 to 01Jul2020, the event date is then on 01Jun. The first event date is the start date for the schedule, but the last event date is not the end date of the schedule; it is the date that immediately precedes the end date in the schedule: 01May2021.

Similar to macros, the manipulation takes place before the scripting library is invoked. The user writes script 2, but the scripting library receives script 1 on its highest level. The transformation is trivial: all scripts on a schedule are copied for each event date in the schedule. Whenever that results in multiple scripts with the same event date, they are merged into a single script for that event date. At that point, events are strings and so by merged we really mean concatenated. In our barrier script, the last date on the monitoring schedule corresponds to the date of the payment of the payoff, resulting in two scripts for 01Jun2021:

```
if spot() > BARRIER then vAlive = 0 endIf
```

and

```
opt pays vAlive * CALL( spot(), STRIKE)
```

With our concatenation rule, including adding a space between strings, the full event for 01Jun2021 becomes

```
if spot() > BARRIER then vAlive = 0 endIf
opt pays vAlive * CALL( spot(), STRIKE)
```

This is a multi-statement event that our parser is designed to support.

Of course, we need a generator to produce the actual schedule, that is, all the periods with their fixing, start, and end dates. This work is typically delegated to a financial date library. It is easy to write a simple date library, but it is time consuming to develop a correct one, especially with a proper treatment of holiday calendars and non-business-day adjustments. Financial institutions generally have decent date libraries readily available. For academic work, it is typically acceptable to roll all events dates to weekdays.

We discussed setting event dates at the start or the end of periods in the schedule. It is also convenient, especially for fixed income products to be able set the fixing date some business days prior to the start or end of the period. In the typical floating leg of a swap, Libor is fixed two business days prior to the start of the period, and paid at the end of period, multiplied by the day count, also called *coverage*, between the start and the end dates of the period. Coverage is computed according to a convention, called *day basis*, the most frequent being act/360 (actual number of days between start and end, divided by 360), act/365, act/act (divide by the actual number of days in that year) and 30/360 (count 30 exactly for each month, divide by 360). act/360 is also called *money market basis* and is typically used for floating coupons. 30/360 is also called *bond basis* and typically is used for fixed coupons.

It is also convenient to predefine in the language two special macros for schedules: *StartPeriod* and *EndPeriod*, the start and end dates of the current period in the schedule. For instance, we will need to develop a function *cvg* that computes the coverage between two dates in a given basis. This function takes a start date, an end date and a basis, and returns the coverage. With the special macros *StartPeriod* and *EndPeriod*, the following script:

start: 01Jun2020 end: 01Jun2021 freq: 3m fixing: start-2bd	fixLeg pays CPN * cvg(StartPeriod, EndPeriod, act/360)

resolves into (disregarding weekends and holidays):

30May2020	fixLeg pays CPN * cvg(01Jun2020, 01Sep2020, act/360)
30Aug2020	fixLeg pays CPN * cvg(01Sep2020, 01Dec2020, act/360)
29Nov2020	fixLeg pays CPN * cvg(01Dec2020, 01Mar2021, act/360)
28Feb2021	fixLeg pays CPN * cvg(01Mar2021, 01Jun2021, act/360)

So, the cash-flows are repeated over the event dates in the schedule, but the definition of *StartPeriod* and *EndPeriod* changes from period to period. Since *cvg* takes different dates as inputs on each period, that script could not be put on a schedule without these special keywords. In addition, the use of *StartPeriod* and *EndPeriod* instead of actual dates makes the script more readable and expressive.

We may as well implement a three special macro *FixingDate*, which refers to the event date for the event.

It is also beneficial to implement schedules in such a way that they accept definitions as parameters. This helps build template scripts. For instance, we can build a template for all up and out barrier options:

STRIKE	100
BARRIER	120
PAYOFF(S, K)	max(0, S-K)
MATURITY	01Jun2021
BARSTART	01Jun2020
BAREND	01Jun2021
BARFREQ	1m
start: BARSTART end: BAREND freq: BARFREQ fixing: end	if spot() > BARRIER then vAlive = 0 endIf
MATURITY	opt pays vAlive * PAYOFF(spot(), STRIKE)

One last suggested improvement is the ability to use either dates or tenors for the end date of the schedule. For example:

01Jun2020	vAlive = 1
start: 01Jun2020 end: 1y freq: 1m fixing: end	if spot() > BARRIER then vAlive = 0 endIf
01Jun2021	opt pays vAlive * PAYOFF(spot(), STRIKE)

Support for schedules is a reasonably easy development of another functionality that is entirely conducted *before parsing*. It makes a significant difference in the practicality of the scripting language. It allows writing script templates for barriers, cliquets, Asian options, and so on in a compact, expressive manner, and neatly supports fixed income products, whose coupons typically lie on a schedule. For these, of course, we also need support for interest rates in the language, which we discuss in part III.

CHAPTER 9

Support for Dates

The function *cvg*, introduced in chapter 8, is unlike any function we have implemented so far in our scripting language. Its arguments are not expressions that evaluate a number. The arguments are two dates and a basis.

When our basic parser (precisely its method *parseVarConstFunc()*, which deals with functions) hits the name of a function, like *log* or *sqrt*, it calls *parseFuncArg()*, which sends all arguments to *parseExpr()* and *parseExpr*() only supports numerical expressions. In particular, it would parse the basis "act/360" as the division of the variable *act* by the constant 360. It should be clear that a function like *cvg* requires a special treatment in *parseVarConstFunc()*. Instead of sending its arguments to *parseExpr()*, the first two arguments should be sent to a new function *parseDateExpr()* and the three to another new function that may be named *parseBasis()*. The same applies to other functions that take for arguments dates or things like basis, tenors, or frequencies. We will meet a few of these in the next section, so we now discuss support for dates in the language and illustrate our discussion with the example of *cvg*.

We start with *parseBasis*(). All it needs to do is to read the basis code from the tokens[1] so that the basis code is stored on the *cvg* node. There is no need to create a constant string node type to hold the argument as a string, or a basis node type to hold the basis code on a separate node. The basis argument cannot be anything else than a string that represents a valid basis code, and that interpretation can be conducted on the fly at parsing time in *parseBasis*(). So, *parseBasis*() reads the basis code from the tokens of the third argument and returns it so it can be stored on the node for the function. The exact same pattern applies to functions like *parseTenor*() and *parseFreq*().

[1]The event string is tokenized before parsing, so that something like "act/360" becomes "act," "/," "360," which simplifies the work of *parseBasis*().

Of course the same can be done with the date arguments, but with a twist. It is useful to allow date expressions, like "01Jun2021+1y" or "StartPeriod-3m." Note that these are not expressions in the same sense that "Spot()-100" is an expression. Dates cannot be stochastic. They are not allowed to depend on scenarios and they cannot be assigned to variables. Thus, any date expression is necessarily a constant expression, one that is evaluated before simulations take place. We could consider building a dedicated date processor to precompute the results of date expressions, but that would be overkill. We may assume that all date expressions are of type *date*, *date* + *tenor*, or *date* – *tenor* (support for expressions like "01Jun2021+1y-3m" does not add much value). Given that this is the case, it is easy to evaluate the expression directly at parsing time and store its result on the node for the function. In this case, *parseDateExpr*() does not only parse the date expression but also actually evaluates it.

As a general rule, it is our philosophy to limit parsing to parsing and use dedicated visitors for anything else. To some extent, that would be a cleaner development. But in this particular case, we believe that adding/subtracting a fixed number of years/months/weeks/days to a constant date is not worth the effort of building new node types and new pre-processors. To do this directly at parsing time is a quick and easy development.

This means that the nodes for the arguments of *cvg* are never built. The arguments are directly evaluated at parsing time and stored on the node for the function. In fact, the result of *cvg* will therefore also be known at parsing time and can be stored on the *cvg* node.

Wait a minute. The function *cvg* does not take any other arguments. From the moment that all its arguments are known, *cvg* can be evaluated straight away. So it can be directly evaluated at parsing time and resolved into a constant node. We don't need to design a *cvg* node. This is however specific to *cvg*. Other functions that take dates or string as arguments, generally also take additional non-deterministic arguments, and so they may not be evaluated at parsing or pre-processing time. However, their date and string arguments should still evaluate at parsing time by calls to *parseDateExpr*() and the likes of *parseBasis*() so that the results for these arguments are stored on the node for the function for an efficient subsequent evaluation of the function at run time.

In summary, support for dates is another light improvement. This one does not take place before parsing, but during parsing, yet it still doesn't involve the creation of nodes, the development of visitors or any change in our code outside of a minor development in the parser. Date expressions leave parsing as constant dates stored on the node of the function; the likes of basis, frequencies, and tenors leave parsing as codes also stored on the function node. The function *cvg* is even evaluated at parsing time and leaves parsing as a constant node that holds its result.

Note that it is the key feature of deterministic dates that makes the support for dates a relatively light development. In turn, the deterministic date feature will also be key to the support and efficient implementation of interest rates and in particular stochastic interest rates, the latter because many stochastic interest rate models do not have closed form solutions for the discount factors.

CHAPTER 10

Predefined Schedules and Functions

Predefined schedules are very convenient whenever some feature of a transaction, say a barrier for an option, or the notional for a swap, is meant to change on a schedule, which is very frequent in financial transactions. For example, consider a 3y barrier option where the barrier is 110 the first year, 120 the second year, and 130 the third year. We monitor the barrier monthly. We could write the script as follows:

01Jun2020	vAlive = 1
start: 01Jul2020 end: 01Jun2021 freq: 1m fixing: end	if spot() > 110 then vAlive = 0 endIf
start: 01Jul2021 end: 01Jun2022 freq: 1m fixing: end	if spot() > 120 then vAlive = 0 endIf
start: 01Jul2022 end: 01Jun2023 freq: 1m fixing: end	if spot() > 130 then vAlive = 0 endIf
01Jun2023	opt pays vAlive * CALL(spot(), STRIKE)

This is not satisfactory: the script is longer than it should be because we monitor the barrier on three different schedules. What if the barrier changed every month? In addition, the script is not expressive and does not make it obvious at first sight that the barrier is on a schedule.

Predefined schedules resolve such situations. First, we create a schedule object (consisting of a collection of dates and corresponding values, called *knots*, stored for instance in an STL map, together with an interpolation/extrapolation method), give it a name "barrier," specify its knots:

01Jun2021	110
01Jun2022	120
01Jun2023	130

We also specify that the interpolation is piece-wise constant, left continuous, that the extrapolation is constant on both sides, and pass the schedule to the scripting library along with the script. The script can now be written in a crisp, expressive form:

01Jun2020	vAlive = 1
start: 01Jul2020 end: 01Jul2023 freq: 1m fixing: end	if spot() > barrier(EndPeriod) then vAlive = 0 endIf
01Jun2023	opt pays vAlive * CALL(spot(), STRIKE)

The parser identifies the "barrier" token as a predefined schedule, parse its argument as a date expression that is evaluated at parsing time, access the schedule object by name, interpolate the result, and put it on a constant node.

Hence, there is still no need to design new node types or modify visitors. This is another light development that considerably simplifies the scripting of any transaction with schedules, which would otherwise result in unnecessarily long and clumsy scripts.

As with the macros that we considered earlier: if we wish to differentiate values with respect to the schedule knot levels using AAD, we need to create a dedicated schedule node.

Note that a predefined schedule is in fact nothing more or less than a *user-defined function*, in which the argument happens to be a date. We can easily define it in exactly the same way as user-defined functions of numbers. These are less frequently used than the schedules, yet they occasionally

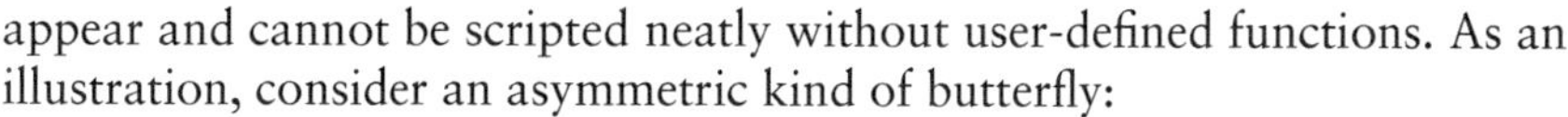
appear and cannot be scripted neatly without user-defined functions. As an illustration, consider an asymmetric kind of butterfly:

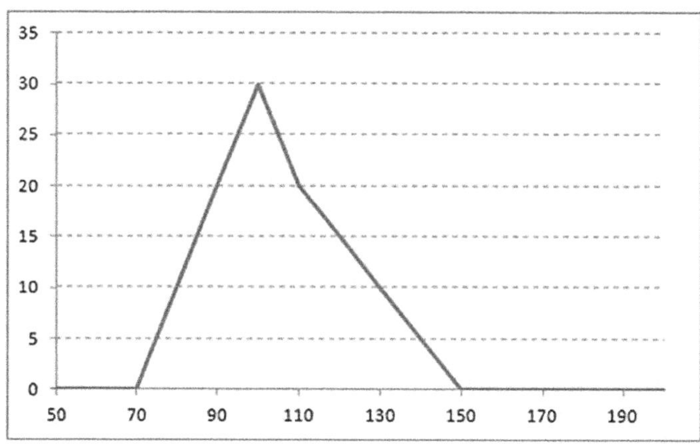

The script for this payoff would involve an awkward set of nested *if*s. Instead, we create a *user-defined function* object, call it "butterfly," specify its knots:

70	0
100	30
110	20
150	0

We also specify that the interpolation is linear and that the extrapolation is constant on both sides, and pass the function (interpolation object) to our scripting library along with the script. The script is now written in a crisp, expressive form:

```
01Jun2021      opt pays butterfly( spot())
```

Contrarily to predefined schedules, which argument is a date and therefore has to be a constant, user-defined functions arguments are full expression trees. Hence, user-defined functions cannot be evaluated at parsing or pre-processing time. We must create a node type for user-defined functions. This node holds a reference to the related interpolation object, or a copy of the object itself. This can be arranged at parsing time. The visitor for the user function node in the evaluator evaluates the argument, interpolates it on the function object, and puts the result on the evaluation stack.

CHAPTER 11

Support for Vectors

The improvements considered so far are high level in the sense that they consist essentially of the pre-treatment of the scripts with very little changes in the scripting library. We now look into an improvement that does require a deeper development: the support for *vectors*.

11.1 BASIC FUNCTIONALITY

Vectors simplify scripts in a vast number of cases of practical relevance. Let us start with the example of an Asian option, which is an option on the average spot price across a number of fixings. We could write the script as follows:

01Jun2020	vAverage = spot() vN = 1
start: 01Jul2020 end: 01Jun2021 freq: 1m fixing: end	vAverage = vAverage + spot() vN = vN + 1
01Jun2021	opt pays CALL(vAverage / vN , STRIKE)

That script is somewhat low level. With vectors, we can write it in the clearer and more expressive (some would call it *functional*) form:

01Jun2020	append (vFixings, spot())
start: 01Jul2020 end: 1y freq: 1m fixing: end	append (vFixings, spot())
01Jun2021	opt pays CALL(average(vFixings) , STRIKE)

Let us first discuss a basic implementation of vectors with only the function *append* that adds an entry to a vector, and some functions like *average*, *maximum*, *minimum*, and so on that take a vector as an argument and return a numerical result based on its size and entries.

We need a new leaf node type for vectors, similar to the variable node, that holds the name of the vector and its index among all the vectors in the script. We also need nodes for the function *append* and the likes of *average*. We must add support for these functions in the parser. The parsing of these functions differs from the parsing of other functions in that their vector argument is not an expression but a name and nothing else. We generally find that the support for functions and operators that *return* vectors is not that useful and we do not implement it. Hence, a vector argument has to be named vector, and that name is the name of the vector, for which a vector node must be created as a child to the function node.

At processing time, we must count and index vectors like we do for variables. We may either extend the variable indexer or develop a similar dedicated vector indexer. At the same time, our indexer should compute the *maximum* size of each vector so we can *reserve* their memory before simulations take place to ensure that no costly allocations occur during simulations.

Finally, we need to develop support for vectors in the evaluator. We must write the evaluator's visitors to vector and vector functions nodes, but, more fundamentally, we must extend the *state* of the evaluator to include vectors along with variables. Vectors live in the evaluator as a collection of pointers to STL vectors. These STL vectors have memory reserved for their maximum size before simulations take place. They are resized to 0 (but not de-allocated) before each simulation. The visitor to the vector node accesses the corresponding STL vector by a pointer that lives in the evaluator (just like we access the variables by pointer when we write into them—vectors are always accessed by reference, even for reading) so parent nodes know where to find the target vector. The visitor to the *append* node makes a call to push_back() on the corresponding STL vector (accessed by its index), which increases its size but does not require allocation provided memory was previously correctly reserved. The visitors to the likes of *average* read access that vector, conduct the necessary calculations from its state (size and entries), and push the result onto the stack.

Just like variables, vectors live in the evaluator's state and are accessed by their index, set by the indexer. The only difference is that the maximum size of vectors is also pre-computed and reserved at processing time.

11.2 ADVANCED FUNCTIONALITY

There is another use for vectors. Say we are scripting multiple European options in a single script. We could do that as follows:

STRIKE1	100
STRIKE2	120
STRIKE3	150
MATURITY	01Jun2021
CALL	max(0, S-K)
MATURITY	s = spot() opt1 pays CALL(s, STRIKE1) opt2 pays CALL (s, STRIKE2) opt3 pays CALL (s, STRIKE3)

Such scripts may end up a maintenance nightmare with more options, so, clearly, we would rather script this with vectors. The script would look something like:

NOPT	3
MATURITY	s = spot() loop (ii, 1, NOPT) opt[ii] pays CALL(s, STRIKES[ii]) endLoop

using the *predefined* vector STRIKES:

100
120
150

We used a few new features here:

Random access to vector entries for writing and reading entries with the C/C++ style [] operator counting from 0: *opt*[*ii*]. We must extend assignment and pays to write into vector entries just like variables. We also need a "vector entry" node type so vector entries may be read as part of expressions (again, same as variables). Note that we don't need full-fledged random access, where the index could be itself a variable or an expression. Constant indices are good enough for us. That means that *v*[*expression*] or *v*[*variable*] is forbidden. Only *v*[*constant*] is allowed, which both simplifies implementation and reduces overhead at run time.

Predefined vectors like STRIKES, similar to predefined schedules from chapter 10.

Loops as in "loop(ii, firstIndex, lastIndex) ... endLoop." We may restrict support to constant *startIndex* and *lastIndex*. It is not necessary either to support nested loops. Note that *ii* is not a variable but a keyword that stands for the current index in the loop.

Use of vectors to reference products, not only ancillary variables.

The implementation of these functionalities requires work on various parts of the code, specifically, new node types, support in the parser, processing, and evaluation.

11.2.1 New node types

We need node types for vector entries. These nodes must store the name and index of the vector and the index of the entry. Because of loops, the index of the entry (*ii* in *opt*[*ii*]) must be an argument and not just a number stored on the node. Vector entries are otherwise identical to vector functions *average*, *maximum*, and so on. Indeed, *v*[*ii*] is "syntactic sugar" for something like *entry*(*v*, *ii*).

We also need node types for predefined vector entries. Contrarily to predefined schedules, these cannot be resolved at parsing time because of loops. Predefined vector entries should store the predefined vector objects, preferably by reference, and refer to the index of the entry as an argument.

Finally, we need a loop node type, which holds its inner statements as arguments, and stores its first and last index on the node. We also need a counter node type for the counter *ii*.

11.2.2 Support in the parser

The parser needs to be extended so it correctly builds the new node types. These extensions are relatively straightforward. The parsing of loops is similar to the parsing of *if* statements. Hierarchically, the *loop* node sits along with =, *pays*, and *if*.

11.2.3 Processing

The vector indexer must go through vector entry nodes to identify and list vectors and determine their maximum size. When vector entries are accessed in loops, the last index for the counter must be considered. This is not necessary for predefined vectors, but we should still check that we don't attempt to read non-existing entries.

11.2.4 Evaluation

The evaluation of a vector entry starts with the evaluation of the index argument, and after that it is identical to the evaluation of a variable. When a vector entry is on the LHS of an assignment or a payment, the address of the entry on the target STL vector should be set on the evaluator's LHS variable pointer so it is written into. Otherwise, that entry is read and its value is put on the stack.

The evaluation of a predefined vector entry also starts with the evaluation of the index argument, followed by a direct read of the corresponding entry in the vector that is put on the stack.

The evaluation of a *loop* node consists of the repeated evaluation of its arguments, and the counter (that must sit on the evaluator as a new data member) being set to different values every time. Finally, the evaluation of the counter node simply reads that state and puts it onto the stack.

Finally, we want all or some of the vectors involved in the script to be considered as products. This simply means that we return their values from the call to the high-level valuation function.

In the section dedicated to multiple underlying assets, we will also need support for predefined vectors *of strings*. This is a simple development that is best conducted at the same time as the support for vectors of numbers.

An alternative for loops is to *unroll* them at processing time. That means that we clone the whole argument tree as many times as the loop repeats and then traverse the trees to pre-compute the counters. That way we can store entry indices directly on the vector entry nodes. Further, predefined vector entries can be pre-calculated and stored in constant nodes. The loop node becomes a *collector* node that collects statement trees that are executed in sequence at run time. Such implementation may optimize performance.

We have discussed basic improvements that make it easier to use the scripting language. But for real-life production use, we also need support for multiple underlying assets, currencies, and interest rates. These developments are more involved and will be covered in the next section.

PART III

Advanced Improvements

Introduction

The previous section discussed improvements in the structure of the scripting language. Most of the implementation work took place in the parser or even string transformations prior to parsing. Only support for user-defined functions and vectors required (rather light) extensions for nodes and visitors. We did not need to discuss models in any way, and in particular, the communication channels with models were not altered.

In this section, we discuss important missing functionality, which is support for interest rates, multiple underlying assets, multiple currencies, and the Least Squares Method (LSM) for pricing Bermudan options with Monte-Carlo.[1] The implementation of such features does not only require the creation and extension of nodes and visitors but also considerably complicates the communication with the models. So far, we have only communicated the collection of event dates to the models before simulations, and models only communicated back the spot and the numeraire for all event dates during each simulation. With interest rates and multiple underlying assets, when we communicate event dates, we must also specify exactly what it is we will need from the model on those event dates, and that information needs to be figured out ex ante by a dedicated visitor. For maximal performance during simulations, the model needs to write simulated data into prespecified regions of pre-allocated memory so that no allocation occurs at run time. Likewise, the evaluator needs to be able to random access simulated data without any type of lookup. This is achieved by the *indexing* of simulated data, similarly to how we index variables and vectors, although in this case, this affects the model, too.

To correctly implement such extensions takes work. There is considerable implementation effort on the extension of our *SimulData* object (which so far has only held two numbers, the spot and the numeraire), the design of a new *SimulRequest* object to represent the information required from the model on different event dates, the creation of a new visitor to identify that information from the script and index it so it is quickly accessible during scenarios, and accordingly, a profound redesign of our *ScriptModelApi*

[1] See [6], [22], and [18].

wrappers for models to work with the scripting library. We advise readers to familiarize themselves with paragraph 3.7, chapter 3, part I.

Chapter 15 discusses support for LSM in the scripting library, assuming an implementation of LSM is readily available. We refer to [18] for implementation details.

Source code is not provided but implementation is discussed in depth.

CHAPTER 12

Linear Products

Linear products are products that can be priced with discount factors and forwards (prices or rates) only, hence, without knowledge of their future stochastic evolution. Such products include interest rate swaps, forwards on interest rates, equities, foreign exchange and commodities, and cross-currency swaps.

There are many products that essentially are derivatives on linear products, particularly in the space of interest rate derivatives, for example, swaptions and caps, which are respectively options on swaps and strips of options on one-period swaps (deposits). For these products it is generally convenient to have direct or indirect access to discount factors and forwards in the scripting language.

Another feature that we need to support is where the payout is delayed after the fixing date and/or paid in another currency. We need the language to be able to supply forwards and discount factors, either directly or through interest rates or values of swaps or values of a delayed payment.

This chapter summarizes some general and theoretical considerations about the pricing of linear products under different collateral rules and gives a general recipe for efficient support of linear products in scripting languages: processing and indexing.

12.1 INTEREST RATES AND SWAPS

Natural interest rate primitives are discount factors and forward Libor rates. Let $P(t, T)$ be the time t value of a discount factor maturing at time T, that is, the time t present value of one unit of currency paid at time T. And let $L(t, T_1, T_2)$ be the time t discrete forward rate for deposit over the interval $[T_1, T_2]$.

For simplicity, we will here use the academic daycount basis $cvg(T_1, T_2) = T_2 - T_1$ rather than the typical "ACT/360." We will also assume that each interest rate period's end date is the start date of the next interest rate period and that the end date is exactly the payment date as well.

This is never quite the case in practice. Details of this type are important for deployment but not really for the discussion here. All formulas in this section are easily modified to include more realistic day count and date schedule generation.

Conventionally, Libors and discount factors are linked by

$$L\left(t, T_1, T_2\right) = \frac{1}{T_2 - T_1}\left[\frac{P\left(t, T_1\right)}{P\left(t, T_2\right)} - 1\right]$$

This is still true, but it is important to note that it is only the case if the Libor and the product in question are under the same collateral rules, which is typically not the case. Standard interest rate swaps are generally daily collateralized, whereas the forward Libor $L(t, T_1, T_2)$ refers to an uncollateralized interbank deposit between $[T_1, T_2]$.

For the discount factor $P(t, T, I)$ we include one extra argument for the collateral I in $[t, T]$. We can relate P to the instantaneous discount rate for collateral I, $r(\cdot, I)$ by the formula

$$P(t, T, I) = E_t\left[e^{-\int_t^T r(u,I)du}\right]$$

where $E_t[\cdot]$ denotes risk neutral expectation conditional on time t information.

For the forward Libor $L(t, T_1, T_2, I, J)$ we include two extra arguments, I, which specifies the collateral over the period to payment $[t, T_2]$, and J, which specifies the collateral for the underlying interbank deposit over the period $[T_1, T_2]$. We have

$$L\left(T_1, T_1, T_2, I, J\right) = \frac{1}{T_2 - T_1}\left[\frac{1}{P\left(T_1, T_2, J\right)} - 1\right]$$

and

$$\begin{aligned}
L\left(t, T_1, T_2, I, J\right) &= P\left(t, T_2, I\right)^{-1} E_t\left[e^{-\int_t^{T_2} r(u,I)du} L\left(T_1, T_1, T_2, I, J\right)\right] \\
&= \frac{1}{T_2 - T_1}\left[P\left(t, T_2, I\right)^{-1} E_t\left[e^{-\int_t^{T_1} r(u,I)du} \frac{P\left(T_1, T_2, I\right)}{P\left(T_1, T_2, J\right)}\right] - 1\right] \\
&= \frac{1}{T_2 - T_1}\left[\frac{P\left(t, T_1, I\right)}{P\left(t, T_2, I\right)} E_t\left[\frac{e^{-\int_t^{T_1} r(u,I)du}}{P\left(t, T_1, I\right)} \frac{P\left(T_1, T_2, I\right)}{P\left(T_1, T_2, J\right)}\right] - 1\right]
\end{aligned}$$

This shows that the effect of the discount collateral on the Libor forward rate is linked to the covariance between the discount rate $r(\cdot, I)$ and the spread $r(\cdot, I) - r(\cdot, J)$. For moderate maturity the effect of changing collateral on the Libor forward may be modest but not necessarily negligible.

After these considerations, we arrive at the following pricing formula for a floating leg over the periods $\{T_0, \ldots, T_K\}$:

$$B(t) = \sum_{k=0}^{K-1} \left(T_{k+1} - T_k\right) P\left(t, T_{k+1}, I\right) L\left(t, T_k, T_{k+1}, I, J_k\right)$$

where I is the index of the collateral of the swap and J_k is the index of collateral of the rate for period k. For a vanilla interest rate swap we typically have I being the overnight deposit and J_k being the Libor deposit over $[T_k, T_{k+1}]$.

For a fixed leg over the periods $\{U_0, \ldots, U_N\}$, an annuity, we have

$$A(t) = \sum_{n=0}^{N-1} \left(U_{n+1} - U_n\right) P\left(t, U_{n+1}, I\right)$$

with the par swap rate being given as $S(t) = B(t)/A(t)$.

We see that support for linear interest rate products requires communication between model and script evaluator of discount factors with four dimensions (fix date, maturity date, collateral, currency) and forward Libor rates with six dimensions (fix date, start date, end date, discount collateral, rate collateral, currency).

12.2 EQUITIES, FOREIGN EXCHANGE, AND COMMODITIES

Equity, foreign exchange, and commodities forwards are generally given by

$$F(t, T, I) = P(t, T, I)^{-1} E_t \left[e^{-\int_t^T r(u,I)} S(T)\right]$$

where t is the fix date, T is the maturity date, I specifies the collateral of the forward, and $S(t)$ is the spot price at time t.

We see that supporting equities, foreign exchange, and commodities will require four dimensions (fix date, maturity date, collateral, extra) where the "extra" is additional information to specify the index that we consider, that is, the name of the stock for the equity, the currency pair for foreign exchange, and the type of commodity.

12.3 LINEAR MODEL IMPLEMENTATION

The most efficient way of implementing support for linear products is if the model is only required to produce exactly the discount factors and forwards that are needed at each event date, and if any node that requires a forward and discount factor knows exactly where in memory these quantities sit.

This avoids expensive calculations, for example, in the case where the discount factors are not known in closed form for the particular interest rate model used, and it avoids costly lookups required when the node needs to search for a specific maturity discount factor or forward in an ordered table.

This thus requires two visitors:

- A *processor* that for each event date identifies exactly what maturity discount factors and forwards are needed for all the nodes in the expression trees
- An *indexer* that for each node identifies where in the aggregated collection of market data the required piece of information is located

Use of these visitors runs in two steps: first, the processor visits all nodes to gather information about what linear price data is required at what date. These data are ordered for subsequent lookup.

Second, the indexer visits the expression trees and for each node it identifies exactly at what slot in memory the required data are located. The position is recorded on each node and this is used during evaluators for direct access. This way, lookup is performed only once, at the indexer stage, and not by the evaluator.

In the processor stage, information is aggregated from the nodes of the expression trees. In the indexer stage, information is set on the nodes of the expression trees.

CHAPTER 13

Fixed Income Instruments

In this chapter we draw on the theoretical considerations of chapter 12 and implement support for linear interest rate products in a number of steps.

13.1 DELAYED PAYMENTS

Let us start with the support for delayed payments. We already provided basic support for the discounting of payments using the *pays* keyword. However, as the language stands, we can only deal with payments that occur immediately on the fixing date. The floating leg of a swap typically fixes Libor (2b before) the start date but only pays it on the end date of each period. Hence, the following script is wrong:

```
start: 01Jun2020        floatLeg pays
end: 01Jun2030       libor( StartPeriod, EndPeriod, act/360, L3M)
freq: 3m             * cvg( StartPeriod, EndPeriod, act/360)
fixing: end
```

because the Libor for each period is fixed on (2b before) start, but paid at the end. If we set the event date on the start, we fix at the right time but pay too early. If we set it on the end, we pay at the right time but fix too late. For one coupon, we could script something like

```
FIXDATE     vLibor = Libor( STARTDATE, ENDDATE, BASIS, IDX)
PAYDATE        floatLeg pays vLibor
            * cvg( STARTDATE, ENDDATE, BASIS)
```

but this is clumsy and it will not work on a schedule. The correct solution is to effect the payment on the fixing date but discount it for the delayed payment:

start: 01Jun2020	floatLeg pays
end: 01Jun2030	libor(StartPeriod, EndPeriod, act/360, L3M)
freq: 3m	* cvg(StartPeriod, EndPeriod, act/360)
fixing: start	* df(EndPeriod)

where *df* is the discount factor simulated on the fixing date for the payment date.

13.2 DISCOUNT FACTORS

We now discuss the support for discount factors, which is implementation around the *df* keyword. We introduce the crucial notion of *indexing* (already introduced for variables in section I, chapter 3, paragraph 3.3, but considerably extended here). We explain the developments in detail in the case of discount factors, because the exact same patterns apply for everything that comes next, especially Libors and other rate fixings, as well as fixings for multiple underlying assets and multiple currencies.

In order to support this new keyword *df*, we need a new leaf node for discount factors. This is a node that accesses simulated data, like *spot*, but this time with a maturity parameter. We also need to identify the keyword *df* in our parser so the parser creates the *df* node, evaluates the maturity date, and stores it on the node. After parsing, the *df* node is a leaf that knows its maturity.

Indexing takes place at processing time and it works as follows when the indexer visits a *df* node. Before simulations take place, the scripting library makes a *request* that the model provides a discount factor of the specified maturity on the fixing date. During simulations, the model will deliver the simulated discount factor and the evaluator will read it and put its value onto the stack.

First, we must extend our *SimulData* structure to include discount factors along with the *spot* and the *numeraire*. How many discount factors do we need? Maybe another statement on the same event date requests another discount factor? Maybe that discount factor is of a different maturity? The *SimulData* object must really hold a *vector* of discounts:

```
template <class T>
struct SimulData
{
  T   spot;
  T   numeraire;
  vector<T> discounts;
};
```

The model will deliver the simulated discounts into that vector. It is best allocation takes place before simulations. In addition, the model must know in advance what discounts of what maturities are needed for different event dates. And the evaluator needs to know where it can random access them. We need a dedicated visitor to figure all that before simulations start. We call this visitor the *simulated data processor* or simply *processor*.

13.3 THE SIMULATED DATA PROCESSOR

The processor visits all *df* nodes before simulations take place and collects for every event date:

- The number of different discount factors needed (that is, discounts with different maturities) so the vector in *SimulData* can be pre-allocated
- The maturities of all discount factors needed so they may be *indexed* in such a way that the model knows them in advance and may set itself up accordingly and the evaluator knows where to find them

How exactly do the model and the evaluator agree on that? We create a new data structure that holds the information about the required simulated data for every event date:

```
struct SimulRequest
{
  vector<Date> discMats;
};
```

This data structure contains, for each event date, the maturities of the discounts that are required for that event. It is constructed by the processor and passed to the model and the evaluator. Hence, we must also extend the API to models:

```
template <class T>
struct ScriptModelApi
{
    virtual void initForScripting(const vector<Date>& eventDates,
        const vector<SimulRequest>& requests) = 0;

    virtual void nextScenario(Scenario<T>& s) = 0;
};
```

Every instantiation of the API for every particular model must set the model up so it produces the requested discounts during simulations.

13.4 INDEXING

Furthermore, at run time, when the model communicates a number of discount factors for some event date, we want the evaluator to pick directly the right one for each *df* node, without the need to look its maturity up in some map or table. Therefore, we need our visitor to also *index* all the required discount factors for every event date and store their indices on the *df* nodes. This may seem confusing, so let us look into an example.

Consider the following (dummy) script:

05Jan2021	X = 100
05Jul2021	Y pays 100 * df(05Jan2022)
05Jan2022	Z1 pays 100 * df(05Jan2023) Z2 pays 100 * df(05Jan2024) Z3 pays 100 * df(05Jan2025) Z4 pays 200 * df(05Jan2025)

After parsing, the discount maturities are stored in the *df* nodes: 05Jan2022 for the 05Jul2021 event; 05Jan2023, 05Jan2024, and 05Jan2025 for the three *df* nodes of the 05Jan2022 event. The processor visits all the *df* nodes and collects the following information:

- No discount factor is needed for 05Jan2021.
- One discount factor is needed for 05Jul2021. Its maturity is 05Jan2022. There is only one, so its index is 0.
- Three discount factors are needed for 05Jan2022. Their maturities are 05Jan2023 (index 0), 05Jan2024 (index 1), and 05Jan2025 (index 2). Even though we have four *df* nodes, the third and fourth node refer to the same maturity; hence, they refer to the same discount factor. We counted three discount factors, so we indexed them 0 to 2.

So as to avoid searching for discount factor maturities at run time, the processor stores the index of each required discount factor on its *df* node. In other terms, we index the discounts in exactly the same way we indexed variables, except that variables are global to the script, whereas discounts are local to an event date. That (per-event date) information about how many discount factors are required, what are their maturities and indices, is all encapsulated in *SimulRequest* :: *discMats*. The size of the vector is the number of required discounts, and *SimulRequest* :: *discMats*[*i*] is the maturity of the discount with index *i*. For instance, in our example, for event date 05Jan2022, we have maturities 05Jan2023 (index 0), 05Jan2024 (index 1), and 05Jan2025 (index 2). Hence $discMats[0] = 05Jan2023$, $discMats[1] = 05Jan2024$, and $discMats[2] = 05Jan2025$.

At run time, in every simulation, the model communicates the requested discounts in the vector *SimulData* :: *discounts* (pre-allocated by the processor), **in the exact same order** as in *SimulRequest* :: *discMats* (which must be ensured by the code in the *nextScenario*() overriding in the API wrapper; hence, contrarily to variable indexing, simulated data indexing requires work not only in the evaluator but also in the API for all models so the same indices are used on both sides). In our example, the model provides no discount factor on 05Jan2021, first discount factor of maturity 05Jan2022 in *SimulData* :: *discounts*[0] for 05Jul2021, and three discount factors, of respective maturities 05Jan2023 (index 0), 05Jan2024 (index 1), and 05Jan2025 (index 2), in the pre-allocated vector of discount factors in the *SimulData* for event date 05Jan2022. When the evaluator visits the *df* nodes, it simply picks the index on that node, reads the corresponding entry from *SimulData* :: *discounts*, and pushes it onto the stack. There is no need to look up maturities on the model or the evaluator side. The resulting performance is equivalent to well-written dedicated pricers and there is no overhead for the use of scripting.

13.5 UPGRADING "PAYS" TO SUPPORT DELAYED PAYMENTS

Dealing with delayed payments with multiplication by discount factors certainly does not qualify as natural language. In addition, in case counterparty default is considered, a payment on the fixing date multiplied by the discount factor is no longer economically identical to a payment on the payment date because a default could hit in between. For these reasons, it is best refraining from using discount factors directly in the script and instead to encapsulate them in the *pays* functionality. That means we use the syntax:

```
product pays amount on maturity
```

as a shortcut for

```
product pays amount * df( maturity)
```

This is more than syntactic sugar. We store delayed payment information in the *pays* node, where it can be retrieved by a visitor in order, for example, to deal with credit risk between the fixing and the payment date.

The *pays* node must be extended to store a payment date (defaulting on the event date in case "on maturity" is omitted) and of course the index for the discount factor; the processor must visit and index *pays* in addition to *df* nodes, and, the evaluation of the *pays* node must multiply the amount by the discount discount picked on the simulated data.

13.6 ANNUITIES

The annuity being a sum of discounts weighted by constant coverages, the annuity node is implemented similarly to *df*, although we also need a schedule processor to generate the internal schedule for annuities and store the schedule dates and coverages on the annuity node. The annuity node is otherwise designed, processed, and evaluated identically to *df*; except, of course, it computes multiple discounts (so it holds a vector of their indices) and sums them up.

13.7 FORWARD DISCOUNT FACTORS

A quantity that in some situations is useful in scripts is a forward discount factor:

$$F(t, T_1, T_2) \equiv \frac{P(t, T_2)}{P(t, T_1)}$$

It is often the case that models can compute forward discounts directly more efficiently than the ratio of the two discounts. For that reason, it is best to only consider forward discounts, *df* taking two parameters, and add code in the parser so *df* with one parameter as in "df(SOMEDATE)" resolves into "df(FixingDate, SOMEDATE)" by default. We need to change *RequestData* (but not *simulData*) and the processor so discounts are indexed by *pairs of maturity dates*:

```
struct SimulRequest
{
  vector<pair<Date, Date>> discMats;
};
```

Note that the fact that discounts are specified, identified, and indexed by a *pair* $<$ *Date*, *Date* $>$ as opposed to a *Date* does not change anything. That *pair* $<$ *Date*, *Date* $>$ is as good as *Date* for that purpose and the *pair* template in C++, just like the more general *tuple* template in C++11, correctly defines equality and ordering out of the box.

13.8 BACK TO EQUITIES

That last comment—that it is better to deal with forward discounts than discounts—also holds for equity, commodity, and forex underlying assets. Let us then improve the interface to single asset models so that scripts can refer not only to their spot, but also *forward* values. This way, we may script an option on a not yet expired forward (also called mid-curve options) as follows:

EXPIRY opt pays CALL (fwd(FWDDATE), K) = 100

As for discounts, it is even best to only communicate forwards, never spots, and extend the parser so that "spot()" still exists but resolves into "fwd(FixingDate)" at parsing time.

How the forward is computed out of interest rates, repo, dividends, convenience yields, and so on is the responsibility of the model.

But it is the responsibility of the processor to compute for each event date how many forwards are required, what their maturity is, index them, and store the index on the *fwd* node (the new *spot* node). During simulations, the model communicates the simulated forwards in exactly the same order so the evaluator can random access them. This is of course identical to our treatment of discount factors, and that means that we extend our *SimulRequest* and *SimulData* objects into:

```
struct SimulRequest
{
   vector<Date>  fwdMats;
   vector<pair<Date, Date>> discMats;
};

template <class T>
struct SimulData
{
  T   numeraire;
  vector<T>  fwds;
  vector<T> discounts;
};
```

13.9 LIBOR AND RATE FIXINGS

Libors, like discounts and forwards, are computed internally by the model. The treatment of Libors is therefore no different from discounts. All Libors for an event date are counted and indexed by the processor, and the maturities for the Libors are stored on the Libor node *along with the index for their curve.* This is actually the only difference: Libors are specified, identified, and indexed with three things: their curve index and their two maturities. Everything is otherwise the same: the extended *SimulRequest* and *SimulData* objects are pre-allocated. The evaluator for the Libor node random accesses the Libor from *SimulData* :: *libors*[*index*] where *index* is the index of the Libor as stored on the Libor node by the processor. Each index identifies a Libor on each event date and uniquely corresponds to a triplet $(curveIdx, T1, T2)$.

```
struct SimulRequest
{
   vector<Date>    fwdMats;
   vector<pair<Date, Date>>  discMats;
   vector<triplet<Idx, Date, Date>> liborMats;
};

template <class T>
struct SimulData
{
  T numeraire;
  vector<T> fwds;
  vector<T> discounts;
  vector<T> libors;
};
```

This type *triplet*, of course, is the exact extension of *pair* to three objects of different types. It is in fact a *typedef* that instantiates the more general standard *tuple* class from the C++11 standard library:

```
#include <tuple>

template <class T1, class T2, class T3>
using triplet = tuple< T1, T2, T3>;
```

The *Idx* class is what identifies curves in the library, which is typically a string.

```
typedef string Idx;
```

Once we have Libors, we can also develop *swapRate* and *swapPV* in exactly the same way we developed annuities when we had discounts.

13.10 SCRIPTS FOR SWAPS AND OPTIONS

That closes our discussion of support for rates. We have all the pieces we need for scripting a receiver swap:

start: STARTDATE end: ENDDATE freq: FLFREQ fixing: start-2b	swap pays -libor(StartPeriod, EndPeriod, FLBASIS, FLIDX) * cvg(StartPeriod, EndPeriod, FLBASIS) on EndPeriod
start: STARTDATE end: ENDDATE freq: FIXFREQ fixing: start-2b	swap pays CPN * cvg(StartPeriod, EndPeriod, FLBASIS) on EndPeriod

We can also make it exotic by capping the floating coupons, for instance:

```
start: STARTDATE    swap pays
end: ENDDATE        -min( CAP,
freq: FLFREQ        libor( StartPeriod, EndPeriod, FLBASIS, FLIDX)
fixing: start-2b    )
                    * cvg( StartPeriod, EndPeriod, FLBASIS)
                    on EndPeriod
```

And, of course, we can change the notional on a predefined schedule, make coupons path-dependent, and so on. For a cap we have the script:

```
start: STARTDATE    cap pays CALL (
end: ENDDATE        libor( StartPeriod, EndPeriod, FLBASIS, FLIDX)
freq: FLFREQ        , STRIKE)
fixing: start-2b    * cvg( StartPeriod, EndPeriod, FLBASIS)
                    on EndPeriod
```

For a swaption, we could use something like:

```
              swaption pays CALL (
EXPIRY-2B     swapPV( START, END, FIXFREQ, FIXBAS,
              STRIKE, FLFREQ, FLBAS, FLIDX)
              , 0)
```

But we advise against that. What this script describes is not the cash-flows of the swaption, but a payoff that is generally deemed economically equivalent. Besides, that only works for standard swaps, for instance, the option on the capped swap could not be scripted this way. This is no longer correct when credit is considered, and important information about cash-flows is lost. That way of valuing swaptions is fine for the calibration of the model, but not in a risk management context. The right way to script an option on a swap would be something like:

```
                    flCpn =
                    libor( StartPeriod, EndPeriod, FLBASIS, FLIDX)
start: STARTDATE    * cvg( StartPeriod, EndPeriod, FLBASIS)
end: ENDDATE        swap pays - flCpn on EndPeriod
freq: FLFREQ        if vExercised = 1
fixing: start-2b
                    then swaption pays -flCpn on EndPeriod
                    endIf
```

start: STARTDATE end: ENDDATE freq: FLFREQ fixing: start-2b	fixCpn = CPN * cvg(StartPeriod, EndPeriod, FLBASIS) swap pays fixCpn on EndPeriod if vExercised = 1 then swaption pays fixCpn on EndPeriod endIf
STARTDATE-2B	if PV(swap) > 0 then vExercised = 1 else vExercised = 0 endIf

The last line of the script (which is chronologically the first) deserves a comment, especially since it looks like we are still missing a piece. It says that if the *PV* of the product *swap* is positive, we set *vExercise* to 1, meaning we exercise, of course; otherwise we set it to 0. Later, the cash-flows of the swaption are all those of the swap, multiplied by *vExercised*, which means that they are only counted when the swaption was exercised, that is, when *PV*(*swap*) is positive on the exercise date.

What is this new *PV* keyword and how can we possibly compute the future PV of a subsequent schedule of cash-flows? This is the raison d'etre of the LSM algorithm for American Monte-Carlo, which is described in detail in a vast number of publications, not least the original paper [22]. Evidently, an implementation of LSM needs to be available in the library. Huge and Savine's recent [18] describes exactly how it is put together for maximum performance. LSM must also be supported in the scripting library, which is the object of chapter 15.

CHAPTER 14

Multiple Underlying Assets

14.1 MULTIPLE ASSETS

Some financial derivatives depend on more than one underlying asset. When this is the case, both the model and the scripting language must support multiple underlying assets. We consider for now multiple underlying assets of equity or commodity type in the same currency.

The model typically consists of a collection of single underlying models, together with a correlation matrix. One scenario consists of a joint path for all underlying assets. We assume that the model and the scripting library agree on a naming convention for underlying assets. If this is not the case, the translation needs to be coded in the *ScriptModelApi*.

The implementation of models is out of scope here; we refer to, for example, Andersen and Piterbarg's [20]. Support in the scripting library is similar to that of Libors, because each basis curve is a named curve just like individual assets are named assets. In practice, this means that the *fwd* and *spot* keywords take an additional parameter that is the code for the asset (typically a string) and should default to some default asset code passed along with the script. So *spot(someCode)* refers to the asset with that name, whereas *spot()* refers to the default. The *fwd* nodes store the code of the asset. This is all conducted at parsing time.

Now, the processor no longer indexes forwards based only on maturity, but also on asset code. Each forward for an event date is specified, identified, and indexed according to its code and maturity. The *SimulRequest* object is extended accordingly:

```
struct SimulRequest
{
   vector<pair<Code, Date>>  fwdMats;
   vector<pair<Date, Date>>  discMats;
   vector<triplet<Idx, Date, Date>> liborMats;
};
```

There is no need to extend *SimulData*.

Together with the support for vectors and predefined vectors, including of strings, introduced in the previous section, we have all we need to script basket options:

NSTOCKS STRIKE	5 0.05
TRADEDATE	loop (0, NSTOCKS) s0[ii] = spot(stocks[ii]) endLoop
EXPIRY	loop (0, NSTOCKS) s1 = spot(stocks[ii]) perf = perf + weights[ii] * (s1 - s0[ii]) / s0[ii] endLoop opt pays max(0, perf - STRIKE)

using the *predefined* vector WEIGHTS:

0.20
0.10
0.15
0.25
0.30

and the predefined vector *of strings* (asset codes) STOCKS:

STOCK1
STOCK2
STOCK3
STOCK4
STOCK5

We can also do a call on one stock with a barrier on another:

PAYSTOCK	STOCK1
BARSTOCK	STOCK2
STRIKE	100
BARRIER	120
PAYOFF(S,K)	max(0, S-K)
MATURITY	01Jun2021
BARSTART	01Jun2020
BAREND	01Jun2021
BARFREQ	1m

BARSTART	vAlive = 1
start: BARSTART end: BAREND freq: BARFREQ fixing: end	if spot(BARSTOCK) > BARRIER then vAlive = 0 endIf
MATURITY	opt pays vAlive * PAYOFF(spot(PAYSTOCK), STRIKE)

14.2 MULTIPLE CURRENCIES

Financial transactions that depend on assets and rates from multiple currencies require a special class of models generally called *quanto models* or *hybrid models*. They are of course again a collection of correlated single asset or yield curve models brought together, but this is significantly more complicated with multiple currencies. First, the model must acknowledge which currency is its *domestic* currency, and all assets and rates denominated in a different currency carry an additional term in their risk-neutral drift, which corresponds to their covariance with the forex to the domestic currency and is called *quanto adjustment*. Second, the global correlation matrix takes hard work especially when the individual yield curve models for each currency are themselves multi-factor. Third, the (potentially stochastic) rates are (part of) the risk-neutral drifts of the assets denominated in that currency, and the RN drift of a forex includes the difference of domestic and foreign rates, where at least one of the rates is itself subject to quanto. Finally, these models are notoriously hard to calibrate. A vast amount of effort had been put in the development of such models in the early 2000s among academicians and Wall Street practitioners in order to correctly risk manage hybrid products like Callable Reverse Power Duals. Nowadays, these models are reused in the context of xVA, which is an even more complicated option, but also depends on multiple assets belonging to different class and currencies.

We assume that such a model is available to us and is able to provide arbitrage-free simulated values for the numeraire (in the domestic currency), together with asset prices, discounts, and Libors belonging to various currencies. A script also needs a notion of what is its domestic currency, ideally the same as the model's, which is the currency in which all payments are made unless explicitly stated otherwise. This is also the currency in which the evaluation is conducted, hence the one in which final values and risk sensitivities are expressed. Further, we assume that the model knows in what currency assets and Libors are denominated and is able to find that information from the asset codes and Libor indices. If this is not the case, some lookup may be needed in the model's API wrapper. The scripting library does not need to know all that. Discount factors, however, don't have an index, so for a discount factor denominated in any other currency than domestic, the currency must be explicitly provided as an argument. For all intents and purposes, this means that the syntax of *df* is always:

```
df( T1 , T2, CCY)
```

where CCY is resolved to the domestic currency at parsing time if omitted. Therefore, discounts need to be indexed by currency as well as start and end time, and the *SimulRequest* object is extended accordingly:

```
struct SimulRequest
{
    vector<pair<Code, Date>>  fwdMats;
    vector<triplet<Ccy, Date, Date>> discMats;
    vector<triplet<Idx, Date, Date>> liborMats;
};
```

(Ccy is generally a typedef for string or some enum.) We identify, specify, and index discounts based on their coordinates in this 3D space of currencies and start and end dates. There is no need to extend *SimulData*.

That takes care of equity, commodity, and rate assets, but forex needs further exploration. For clarity and because a special treatment is needed, we advise separating forex from equity and commodity assets, and using a specialized keyword, nodes, and entries in *SimulRequest* and *SimulData* for forex instead of reusing *spot* and *fwd*.

The keyword to access a forex for/dom is:

```
fx( dom, for, T)
```

where *T* is maturity date of the forward (resolved to *FixingDate* at parsing time if omitted) so we can access a forward or a spot forex. So far, this is very similar to equity/commodity forwards. But we can also develop a very convenient syntax for payments in currencies other than domestic and build it into the *pays* keyword:

```
product pays amount on payDate in payCcy
```

which of course is sugar for

```
product pays amount * fx( domCcy, payCcy, payDate) on payDate
```

and one that makes scripts more natural, expressive, easier to write, and less prone to errors.

How does that work with the model? Say the model includes four currencies: EUR, USD, GBP, and JPY. It is clear that model simulates three forexes. But which ones? Say the domestic currency is GBP. It might be tempting to simulate foreign currencies in terms of the domestic: EUR/GBP, USD/GBP, and JPY/GBP. But those dynamics may be hard to calibrate. Market data are typically available for EUR/USD, GBP/USD, and USD/JPY. Anyhow, the model simulates three forexes of its choice, but the script may request up to 12 different forexes: EUR/USD, USD/EUR, EUR/GBP, GBP/EUR, EUR/JPY, JPY/EUR, USD/GBP, GBP/USD, USD/JPY, JPY/USD, GBP/JPY, and JPY/GBP. The model must have the ability to provide any pair as a product/ratio of the three it simulates. Further, the model must not waste time figuring that out at run time, and must set itself up by indexing the requested forexes in terms of the simulated forexes before the simulation starts. It is not necessarily the case that all available implementations of multi-currency models come with this ability. When it is not the case, that logic must be coded in the model's *ScriptModelApi* wrapper. Anyhow, this is all the responsibility of the model. The scripting library just needs to request and receive all pairs; how the model manages that is not a scripting concern.

The node for the forex must store the domestic and foreign currency codes, the forward maturity, and the index of the forex. The pays node also stores that data. The processor indexes forexes in accordance with their domestic and foreign currencies, and forward maturity dates, from its visits to *fx* and *pays* nodes.

```
struct SimulRequest
{
    vector<pair<Code, Date>>  fwdMats;
    vector<triplet<Ccy,Ccy, Date>> fxMats;
    vector<triplet<Ccy, Date, Date>> discMats;
    vector<triplet<Idx, Date, Date>> liborMats;
};
```

Of course, we also need forexes in SimulData:

```
template <class T>
struct SimulData
{
  T numeraire;
  vector<T> fwds;
  vector<T> fxs;
  vector<T> discounts;
  vector<T> libors;
};
```

In summary, scripts request forexes $(ccy1, ccy2, T)$ either directly or as part of a payment. The processor indexes all forexes according to their *ccy*1, *ccy*2, and *T*, and sets the corresponding indices on the *fx* and *pays* nodes. The scripting library communicates that information to the model through *SimulRequest* objects together with event dates. The model sets itself up to provide all the necessary forexes quickly at run time; in particular, it precomputes and preindexes how to get these out of the forexes it actually simulates. During each simulation, the model conducts the computation and communicates the requested forexes as part of the scenario. The evaluator for the *fx* and *pays* nodes reads the information in with a random access into *SimulData* :: *fxs*[*index*]. The required additional model logic may be coded in its *ScriptModelApi* wrapper.

CHAPTER 15

American Monte-Carlo

We need support for handling early exercise features in our scripting language. This is necessary for European and Bermudan options as well as for calculating xVA and regulatory quantities.

Traditionally, solving early exercise problems was done with backward recursive methods such as finite difference methods. However, in the late 1990s and early 2000s a number of methods emerged that allowed the approximate pricing of early exercise options through the use of Monte-Carlo. Of these methods the most versatile and flexible is the Least Squares Method (LSM) by Carriere (1996), publicized by Longstaff and Schwartz (2001). This is the method that we will concentrate on implementing in our scripting library.

15.1 LEAST SQUARES METHOD

The LSM algorithm provides a way of approximating the future PV within Monte-Carlo simulations of subsequent sequences of cash-flows. LSM was initially developed for callable exotics. At the time, interest rate models had grown in complexity and dimension and could no longer be implemented in terms of finite differences or other backward induction algorithms. They had to be implemented with Monte-Carlo simulations, which, as is well known, run forward and simulate the state variables of the model, not the PV of the transactions. But the decision whether to exercise the option at the exercise date depends on the PV at that date of the subsequent cash-flows. The party who holds the option exercises it when the PV of the subsequent cash-flows is positive from that party's point of view. Within a Monte-Carlo simulation, that PV is not available. The PV of a sequence of cash-flows is the expectation of the sum of the discounted cash-flows, conditional to the filtration at the exercise date; hence, it *is* a function of the state variables at the exercise dates and before, because that filtration is generated by these state variables, but that function is not known. To compute it would normally require *nested simulations* (Monte-Carlo within Monte-Carlo), which is obviously inefficient. LSM resolves the problem by the use of regression proxies that approximate the PV on the exercise date as a function of the state variables at that date, more precisely a linear function of regression variables, themselves basis functions of state variables. The regression proxies are produced during an LSM step that occurs before the Monte-Carlo simulations. The LSM step uses some pre-simulations, typically a reduced number of simulations, where the PV at the exercise date of the subsequent cash-flows is regressed onto the value of the regression variables at the exercise date across scenarios. The regression coefficients are stored and reused during the main Monte-Carlo simulations to estimate the PV knowing the simulated regression variables. The efficiency of the LSM algorithm crucially depends on how well the chosen set of basis functions spans the value of the product. It also requires the number of pre-simulations to be sufficiently large to accurately estimate the regression coefficients.

When a transaction is callable on multiple exercise dates, what we call a *Bermudan* option, regression proxies must be computed recursively, last to first, since the PV of the cash-flows subsequent to some exercise date depends on the exercise strategy for the future exercise dates.

The LSM algorithm has been a golden standard for callable transactions in the context of Monte-Carlo since it emerged in the early 2000s. More recently, it has been widely adopted for the calculation of the counterparty value adjustment (CVA) and various regulatory calculations. We remind that a CVA is a complicated option written on the entire set of transactions against some counterparty, or netting set, with payoff:

$$\sum_i (1 - R(T_i))1_{\{T_i \le \tau < T_{i+1}\}} \max(0, V(T_i))$$

where τ is the default time, R the recovery rate, and V_{T_i} is the PV of the netting set at exposure time T_i, that is, the PV at that time of all future cash-flows from all the transactions in the netting set. In the context of CVA, we need an estimate of the future PVs not to make a decision on exercise, but because those PVs are part of the payoff.

LSM is a key part of a modern quantitative library, and we must support it in our scripting language. Otherwise, we could not script callable transactions or CVA and the language would be limited in scope and use.

Importantly for our purpose, LSM, not unlike the scripting library itself, is model agnostic and should be implemented model independently in such a way that it works with any model that is able to generate scenarios for the state variables. In other terms, the model API required for LSM is *almost* the same as the API we developed for scripting. The only difference is that scripting so far does not require that the model exposes its state variables. But LSM does so let us correct this straight away.

We add a boolean indicator in our *SimulRequest* object that indicates that a proxy is computed at that date. That means that the model is requested to produce its state variables at that date:

```
struct SimulRequest
{
   vector<pair<Code, Date>>  fwdMats;
   vector<triplet<Ccy,Ccy, Date>> fxMats;
   vector<triplet<Ccy, Date, Date>> discMats;
   vector<triplet<Idx, Date, Date>> liborMats;
   bool     showSVs;

   SimulRequest() : showSVs( false) {}
};
```

We also need SVs in *SimulData*:

```
template <class T>
struct SimulData
{
  T  numeraire;
  vector<T> fwds;
  vector<T> fxs;
  vector<T> discounts;
  vector<T> libors;
  vector<T> SVs;
};
```

The vector for holding SVs in the *SimulData* object must be pre-allocated. So, after the model sets itself up for the delivery of the required simulated data, it must expose the number of its state variables:

```
template <class T>
struct ScriptModelApi
{
   virtual void initForScripting(const vector<Date>& eventDates,
      const vector<SimulRequest>& requests) = 0;

   virtual size_t numSV() const = 0

   virtual void nextScenario(Scenario<T>& s) = 0;
};
```

In what follows, we assume that an implementation of LSM is available. Among other functionality, the LSM library must provide a function to compute regression variables out of state variables. For example, when LSM is set up to perform linear regressions, the regression variables will be the state variables, plus the unit variable (one that is always worth 1). With quadratic regressions, regression variables are the unit variable, the state variables, and all their pairwise products.

We refer to Huge and Savine[18] for a full discussion of a modern implementation. LSM is run before Monte-Carlo pricing simulations start. Its only purpose is to compute and store the regression coefficients so that later, during the simulations, estimators for the PVs can be calculated as functions of the simulated state variables (as turned into regression variables by a call to the specialized function in the LSM library), which permits for instance to evaluate our swaption script:

<table>
<tr><td>STARTDATE-2B</td><td>if PV(swap) > 0
then vExercised = 1
else vExercised = 0
endIf</td></tr>
<tr><td>start: STARTDATE
end: ENDDATE
freq: FLFREQ
fixing: start-2b</td><td>flCpn =
libor(StartPeriod, EndPeriod, FLBASIS, FLIDX)
* cvg(StartPeriod, EndPeriod, FLBASIS)

swap pays - flCpn on EndPeriod

if vExercised = 1
then swaption pays -flCpn on EndPeriod
endIf</td></tr>
<tr><td>start: STARTDATE
end: ENDDATE
freq: FLFREQ
fixing: start-2b</td><td>fixCpn = CPN
* cvg(StartPeriod, EndPeriod, FLBASIS)

swap pays fixCpn on EndPeriod

if vExercised = 1
then swaption pays fixCpn on EndPeriod
endIf</td></tr>
</table>

A new visitor is needed to count all calls to PV to ascertain at what event dates proxies are needed.

15.2 ONE PROXY

We start the discussion with a single proxy at a single date, as in our swaption script.

During simulations, PV are estimated, in the evaluator's visitor for the PV node, with the formula:

$$PV = \sum_{i=0}^{n} \beta_i X_i$$

where:

- n is the number of regression variables, excluding the unit variable, and must be provided by a function from the LSM library, given the number of state variables, which is exposed by the model.
- X_i are the regression variables simulated in this scenario for that date. They are computed by a function from the LSM library, given the simulated state variables, which are communicated by the model through the *SimulData* object.
- β_i are the regression coefficients that are produced by the LSM algorithm for that proxy (the LSM algorithm runs before Monte-Carlo simulations and does nothing else than computing these). They are stored on the PV node.

Note that during simulations, the PV node does not use its argument, the variable for which it provides the PV. It is however obviously used in the LSM step. That's fine. By construction, the child node to PV is the variable node and holds its index.

During the LSM step, in order to produce the regression coefficients, the LSM library conducts a regression of the PVs of subsequent cash-flows on that variable over the regression variables simulated for that event date. If the LSM library performs a standard regression, that calculation consists of the well-known matrix computation:[1]

$$\beta = (X'X)^{-1}(X'Y)$$

[1] A better alternative is to use a Tikonov-regularized SVD regression, as explained in Huge and Savine [18].

where $'$ denotes transpose, X is the matrix of all regression variables (in columns) in all pre-simulated scenarios (in rows), and Y is the vector of the values of future cash-flows in the corresponding scenarios.

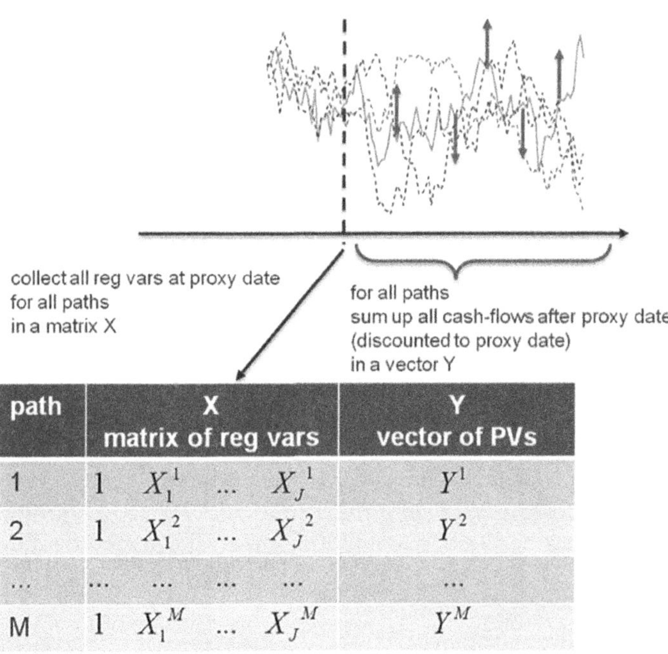

path	X matrix of reg vars				Y vector of PVs
1	1	X_1^1	...	X_J^1	Y^1
2	1	X_1^2	...	X_J^2	Y^2
...	...	...	...	...	...
M	1	X_1^M	...	X_J^M	Y^M

All the LSM library needs to perform the regression is, for each pre-simulated scenario m, the row vector of regression variables X^m and the (scalar) value of future cash-flows Y^m. The LSM library computes X^m as a function of the simulated state variables communicated by the model. To compute Y^m, the LSM library performs the following steps, which require light alteration in the scripting library: an extension of the method *Product* :: *evaluate*() to support *partial* evaluation from some date T_0 to some date T_1 given an evaluator (or only an evaluator state) at T_0.

1. Evaluate all event dates up to and including the proxy date. We only regress cash-flows that happen after the proxy date. Store the value of the regressed variable on the proxy date, denote it V_0, as well as the numeraire N_0.

2. Evaluate all cash-flows after the proxy date. Denote V_1 the final value of the variable.[2]
3. $Y = N_0(V_1 - V_0))$.

15.3 ADDITIONAL REGRESSION VARIABLES

In case the callable transaction is itself path-dependent, for example, if its coupons depend on some cumulated average, then a proxy to its PV that only depends on the state variables may be very inaccurate, because in reality, that PV also depends on the path dependency, in our example, on the average up to the exercise date. This is discussed in Huge and Savine and resolved by treating the path-dependency as an additional regression variable. That means that the LSM library must accept, in each pre-simulated scenario, a collection of regression variables additional to the ones it computes out of the state variables simulated by the model. We must support this functionality in the language. We do so by adding arguments to the PV keyword as follows:

```
PV( regressedVar, regressionVar1, regressionVar2, ... )
```

These arguments are optional and each one is an expression tree. During the LSM step, these arguments are evaluated and their result is sent to the LSM library so it incorporates them among regression variables.

15.4 FEEDBACK AND EXERCISE

One question we have so far ignored is this: in every pre-simulation, we start with the evaluation of all event dates up to *and including* the proxy date. That includes the evaluation of the PV node itself:

STARTDATE-2B	if PV(swap) > 0 then vExercised = 1 else vExercised = 0 endIf

[2]Note that a optimization may be used here. We only need to evaluate the statements that, directly or indirectly, affect the coupons of the regressed variable. For instance, in our script, the regressed variable is swap, and it does not depend on swaption any any way. Hence, the statements for the payments of the swaption don't need to be evaluated. It is actually possible, if not trivial, to identify all the chain of dependencies of variables on variables and variables on statements with a dedicated visitor, and then use that information to optimize LSM by evaluating only events that affect the regressed variable.

Obviously, we don't have regression coefficients yet so we can't compute the proxy. What does the evaluator do when it visits that node during the LSM step? Well, it can't do much more than to return (put on the stack) a default value like 0. This also means that we need to use a different, derived LSM evaluator during the LSM step, one that only overrides the visitor to the PV node and otherwise behaves identically to the evaluator.

Yet, we still have a problem. Our script is written in such a way that the proxy references the underlying swap. But what if we don't script the underlying swap, only the swaption? The following script should work:

STARTDATE-2B	if PV(swaption) > 0 then vExercised = 1 else vExercised = 0 endIf
start: STARTDATE end: ENDDATE freq: FLFREQ fixing: start-2b	flCpn = libor(StartPeriod, EndPeriod, FLBASIS, FLIDX) * cvg(StartPeriod, EndPeriod, FLBASIS) if vExercised = 1 then swaption pays -flCpn on EndPeriod endIf
start: STARTDATE end: ENDDATE freq: FLFREQ fixing: start-2b	fixCpn = CPN * cvg(StartPeriod, EndPeriod, FLBASIS) if vExercised = 1 then swaption pays fixCpn on EndPeriod endIf

In the case of Bermudas, there is no other choice but to write scripts this way, because if we decide to not terminate the Bermuda on some exercise date, we are effectively getting into a Bermuda starting on the following exercise date. Our trick to use a proxy on the underlying (non-callable) transaction only works for one-time callables.

It is easy to see that the new script will fail. During the LSM step, *PV(swaption)* will always return 0, so *vExercised* will always remain 0, so the subsequent cash-flows will always evaluate to 0, *Y* will always evaluate to 0, and the regression will fail.

What we really need to do during the first phase of the LSM evaluation (evaluation of events up to and including the proxy date) is to *ignore exercises*. That is clearly mentioned both in the Longstaff Schwartz paper and in Huge and Savine. The evaluation of Y must be conducted *ignoring exercises up to and including the proxy date.*

This is all good, but how do we identify exercises? The line

```
STARTDATE-2B    if PV( swaption) > 0
                then vExercised = 1
                else vExercised = 0
                endIf
```

exercises "by hand," so the library does not know that this an exercise and that it should be ignored in the LSM step (that means by the derived evaluator), or something other than an exercise, something that must be evaluated. What we need is a syntax for exercises. That way, it is clear that such statement is an exercise, which is a benefit in itself, but, more importantly, it tells the LSM evaluator that some statement is an exercise and should be ignored if its event date is before or on the proxy date.

There is ambivalence when we talk about exercises. With fixed income transactions, an exercise generally cancels the rest of the transaction, not unlike a barrier. The party who has the right to cancel does so when the PV from their point of view is negative. With options, like a call, we generally consider exercise the decision to receive some cash-flow, if positive. For instance, for a standard call option, we exercise when $S - K > 0$, in other terms when the present value of the cash-flow $S - K$ is positive, and in this case we receive $S - K$ upon exercise. But in this case, we could also equivalently consider exercise as the cancellation of a transaction that delivers the cash-flow $S - K$. If the PV is negative, we cancel the transaction and receive nothing. Otherwise, we let it live and receive the cash-flow. It is important to realize that exercises can always be represented both ways. We must choose one. We choose to always treat them as cancellations, and therefore develop the following new statement in our syntax:

```
terminate aliveVar when PV( transaction) < 0
[withRebate rebate [paidOn rebateDate]]
```

which is of course syntactic sugar for:

```
if PV( transaction) < 0 then
aliveVar = 0
transaction pays rebate on rebateDate
endIf
```

And it evaluates in exactly the same way, only now we know, and more importantly visitors know, that we are dealing with an exercise that must be ignored during the LSM step for dates earlier or on the proxy date.

Our new keyword *terminate* should be followed by the name of the variable that is set to 0 on exercise, followed by *when*, followed by a condition for exercise (elementary or otherwise), possibly followed by an expression for a rebate paid on a rebate date in case exercise happens. With this syntax, when the condition is not met, the state of the evaluator is left alone. Hence, it is safe to ignore the whole statement in the LSM step. Note that the whole statement is ignored; the condition is not even evaluated. We can now safely write the following script:

TRADEDATE	vExercised = 1
STARTDATE-2B	terminate vExercised when PV(swaption) < 0
start: STARTDATE end: ENDDATE freq: FLFREQ fixing: start-2b	flCpn = libor(StartPeriod, EndPeriod, FLBASIS, FLIDX) * cvg(StartPeriod, EndPeriod, FLBASIS) if vExercised = 1 then swaption pays -flCpn on EndPeriod endIf
start: STARTDATE end: ENDDATE freq: FLFREQ fixing: start-2b	fixCpn = CPN * cvg(StartPeriod, EndPeriod, FLBASIS) if vExercised = 1 then swaption pays fixCpn on EndPeriod endIf

15.5 MULTIPLE EXERCISE AND RECURSION

That completes supports for LSM in the scripting language in the case of a single proxy, and actually, we also have a working syntax for multiple proxies. Our script easily generalizes to Bermudas; we simply put the exercise on a schedule. In the example, we can exercise on fixed coupon payment dates after two years.

TRADEDATE	vAlive = 1
start: FIRSTEX end: LASTEX freq: EXFREQ fixing: start	terminate vAlive when PV(swaption) < 0
start: STARTDATE end: ENDDATE freq: FLFREQ fixing: start-2b	flCpn = libor(StartPeriod, EndPeriod, FLBASIS, FLIDX) * cvg(StartPeriod, EndPeriod, FLBASIS) if vAlive = 1 then swaption pays -flCpn on EndPeriod endIf
start: STARTDATE end: ENDDATE freq: FLFREQ fixing: start-2b	fixCpn = CPN * cvg(StartPeriod, EndPeriod, FLBASIS) if vAlive = 1 then swaption pays fixCpn on EndPeriod endIf

During the main simulations, *vAlive* is terminated on the first exercise date when the PV is negative. During the LSM step, regression coefficients are computed recursively, last exercise first, first exercise last, as follows.

These steps are performed by the LSM library, making calls to the scripting library for the evaluation of cash-flows. The scripting library uses the derived LSM evaluator.

1. For every pre-simulated scenario m
   (a) Compute the row vector X^m
      i. Access state variables for the proxy date.
      ii. Compute regression variables.
      iii. Evaluate user-defined variables from the script and add them into X^m.
   (b) Evaluate all event dates up to and including that proxy date. *Ignore all exercises.* Store the value of the regressed variable, denote it V_0, and the numeraire N_0.[3]

[3]That part can be optimized so a base simulation ignoring all exercises only occurs once, and the state of the evaluator is stored for all scenarios on all proxy dates.

   (c) Evaluate all cash-flows after the proxy date, *including future exercises*. That is where the recursion lies and why we must compute coefficients for proxies in the reverse chronological order. Denote V_1 the final value of the variable.
   (d) Set $Y^m = N_0(V_1 - V_0)$.
2. Regress Y on X.
3. Store regression coefficients on the PV node.

Note that the LSM evaluator needs some "flick" so that on or before the current proxy date, exercises are ignored and PVs return 0, while after that date, the derived evaluator behaves just like the regular evaluator, evaluating proxies and exercises with the then-known regression coefficients (actually, evaluation on those dates should be delegated to the base evaluator).

These steps are repeated for each proxy, last to first.

Obviously, PV may be used in contexts other than exercise, which makes the language suitable for regulatory calculations like xVA.

Finally, note that the major part of CPU time in the LSM step is spent in the evaluation of scenarios. Evaluations of different paths m (as well as their generation by the model) is an "embarassingly parallel" task that can (and should) be distributed or multi-threaded.

The scripting library is thread safe as long as each thread uses its own copy of the evaluator (both main and LSM).

PART IV

Fuzzy Logic and Risk Sensitivities

Introduction

We have so far covered the valuation of scripted transactions. In this section, we focus on risk sensitivities. These are traditionally computed with finite differences by bumping model parameters one by one and repeating valuation. Recently, the financial community widely adopted adjoint differentiation, a much more efficient differentiation algorithm that computes all derivatives of some calculation code analytically in one single sweep through the reverse sequence of the calculations involved. The valuation code must be instrumented to use a custom number type, in which operators are overloaded so that all mathematical calculations involved in the valuation are recorded on a tape, after all derivative sensitivities are computed with an application of the chain rule backwards through the tape. Hence, adjoint differentiation (or AAD) computes any number of derivatives in constant time. AAD is discussed in numerous textbooks, papers, and presentations, including Giles and Glasserman's [14], who first introduced AAD technology to finance, Uwe Naumann's textbook [24], and the first volume of this series [27].

AAD is not further investigated in this section. In what follows, we focus on the differentiation of discontinuous cash-flows, which is a problem both with finite differences and AAD, resolved in the same manner in both cases. Whether derivative sensitivities are computed with brute force finite differences or adjoint propagation, the differentiation of discontinuous profiles combined with Monte-Carlo simulations produce unstable, unreliable results. This is a well-known problem, and a number of solutions were developed in the academy and the industry. The two main ones are the elegant likelihood ratio method and the effective smoothing method.

One popular means of computing the risk sensitivities of discontinuous payoffs is the so-called likelihood ratio method. Notice that:

$$\frac{\partial}{\partial a_i} E\left[f(\vec{X})\right] = \frac{\partial}{\partial a_i} \int_{\vec{X}} f\left(\vec{X}\right) \varphi\left(\vec{a}, \vec{X}\right) d\vec{X}$$

$$= \int_{\vec{X}} f\left(\vec{X}\right) \varphi_{a_i}\left(\vec{a}, \vec{X}\right) d\vec{X} = \int_{\vec{X}} f\left(\vec{X}\right) \frac{\varphi_{a_i}\left(\vec{a}, \vec{X}\right)}{\varphi\left(\vec{a}, \vec{X}\right)} \varphi\left(\vec{a}, \vec{X}\right) d\vec{X}$$

$$= \int_{\vec{X}} f\left(\vec{X}\right) \left[\log \varphi\left(\vec{a}, \vec{X}\right)\right]_{a_i} \varphi\left(\vec{a}, \vec{X}\right) d\vec{X} = E\left\{f\left(\vec{X}\right) \left[\log \varphi\left(\vec{a}, \vec{X}\right)\right]_{a_i}\right\}$$

so we may differentiate the generally continuous log-likelihood of the risk-neutral distribution of the state variables instead of attempting to differentiate the potentially discontinuous payoff f. The so-called Malliavin weights can be computed path-wise as a function of the derivatives of the drift and the diffusion coefficient of the state variables using known results from Malliavin's calculus. Further, this computation can be made very efficient with reverse adjoint propagation. Unfortunately, the computation of Malliavin weights is *not* model agnostic. It is dependent on the model and must be conducted and maintained for all models and numerical implementations in a library. In addition, depending on the model, the computation of Malliavin weights may or may not be tractable.

For these reasons, likelihood ratio, despite its formal appeal, is not widely applied on trading desks. An alternative, widely applied best practice, is *cash-flow smoothing* (also called *payoff smoothing*), whereby discontinuous cash-flows are replaced by approximate continuous ones. For example, digital cash-flows are approximated by tight call spreads. Knock-out products are approximated by so-called *soft barriers*, whereby only a part of the notional knocks out, depending on how far the barrier index lands into the KO region, the simulation continuing with whatever notional remains. Payoff smoothing requires no work on the models or their implementation, only on the payoffs. Therefore, we do not discuss Malliavin's calculus further in what follows and focus on smoothing and its implementation for scripted products.

One question we want to investigate is how smoothing can be nicely implemented for scripted products in such a way that it doesn't break the natural, readable script syntax or our requirement that it is the raw cash-flows that are meant to be scripted, without any reference to a *solution* like smoothing. This is a problem that, to our knowledge, has not been addressed satisfactorily so far. We offer a solution based on a specific application of the so-called *fuzzy logic*, a framework developed in the 1960s by Lotfi Zadeh for very different purposes. We will demonstrate that standards smoothing techniques for digitals and barriers are particular applications of fuzzy logic. More importantly, we will demonstrate that a *systematic* application of fuzzy logic leads to a general, automated algorithm to identify and correct discontinuities of scripted cash-flows in an elegant, efficient, and practical manner. In particular, our algorithm does not modify scripts in any way; rather, it evaluates the unmodified scripts in a different manner, with a systematic application of fuzzy logic that effectively corrects instabilities and biases in the resulting risk sensitivities.

We identify some limitations to the methodology, but conclude that it performs remarkably well in cases of practical relevance.

We should point out, once again, that an implementation of fuzzy logic would not be feasible if the payoff were a black box. The algorithm needs access to the structure of the cash-flows so it can identify discontinuities and evaluate them with fuzzy logic. In other words, fuzzy logic is another application of the visitor pattern.

No code is provided in the text, written in words, mathematics, and pseudo-code for clarity, but a C++ implementation is available in our online repository.

CHAPTER 16

Risk Sensitivities with Monte-Carlo

16.1 RISK INSTABILITIES

Our simplified autocallable script from part I:

01Jun2020	vRef=spot() vAlive=1

01Jun2021	if spot() > vRef then
	prd=110
	vAlive=0
	endIf

01Jun2022	if spot() > vRef then
	if vAlive=1 then prd=120 endIf
	vAlive=0
	endIf

01Jun2023	if vAlive=1 then
	if spot() > vRef then prd=130 else prd=50 endIf
	endIf

has a discontinuous profile as a function of the underlying asset price at maturity. Provided the transaction survives to maturity, it pays 130 when the spot fixes above the reference and 50 otherwise. This results in a jump by 80 points at the reference price (say 100) in the profile.

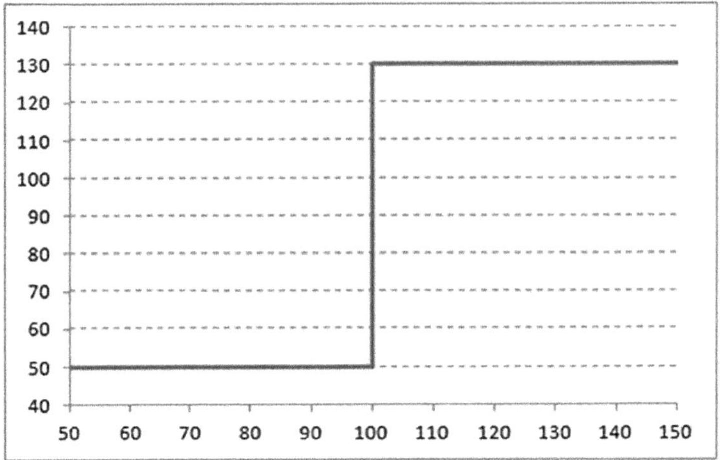

This is the payoff of a so-called digital or binary option, a textbook case of a discontinuous profile. A standard digital pays $1 when the spot terminates above a certain strike, and 0 otherwise, as in the following script (today is 01Jun2020):

01Jun2020	if spot() > 100 then D pays 1 else D pays 0 endIf

Its discontinuous profile as a function of the final spot price is the Heaviside function drawn here:

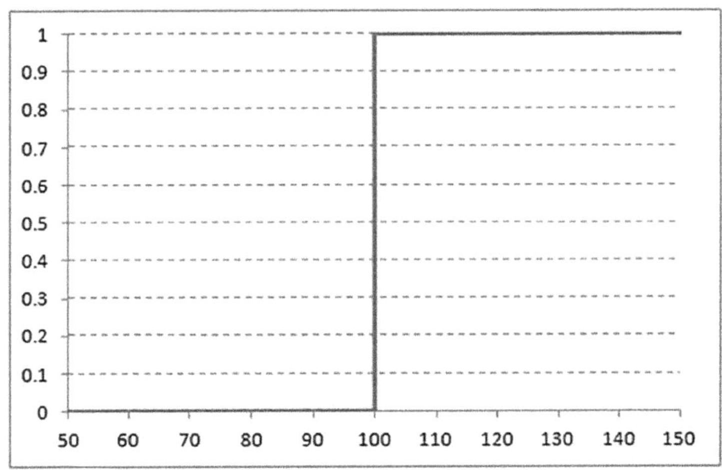

$$D_{100}(S) \equiv 1_{\{S>100\}}$$

A European profile with discontinuities can be written as the sum of a continuous profile and a combination of digitals. Our autocallable's profile at maturity is $50 + 80D_{100}$. A European profile with discontinuities can be written as $c(S) + \sum_{i=0}^{n-1} \alpha_i D_{K_i}(S)$ where c is a continuous function and we have n discontinuities on the points (K_i) with jump sizes (α_i). The PV of such profile is $V_c + \sum_{i=0}^{n-1} \alpha_i V_{D_{K_i}}$ and its sensitivity to some model parameter a (such as the initial spot or its volatility) is $\frac{\partial V_c}{\partial a} + \sum_{i=0}^{n-1} \alpha_i \frac{\partial V_{D_{K_i}}}{\partial a}$.

In this expression, while the Monte-Carlo estimation of $\frac{\partial V_c}{\partial a}$ doesn't raise particular issues, it is the estimation of the $\frac{\partial V_{D_K}}{\partial a}$s that is problematic. The value of a digital (ignoring discounting) is its risk-neutral probability of ending in the money: $V_{D_{K_i}} = \int D_{K_i} d\varphi_a = 1 - \varphi_a\left(K_i\right)$ and is generally differentiable with respect to a, even though the integrand D_{K_i} is discontinuous. The Monte-Carlo estimate of the value from a finite number N of scenarios, however, is $\hat{V}_{D_{K_i}} = \frac{1}{N} \sum_{j=0}^{N-1} D_{K_i}\left[\varphi_a{}^{-1}\left(u_j\right)\right]$, where the $\left(u_j\right)$ belong in $(0,1)$, and this expression is clearly not differentiable in a. As a result, an attempt to compute $\frac{\partial \hat{V}_{D_{K_i}}}{\partial a}$ in this context gives rise to unstable, inaccurate results.

Intuitively, if a is bumped by a small amount, the paths that ended out of the money before the bump still end out of the money, and the paths that ended in the money still end in the money, producing the same value after the bump, hence a sensitivity of 0. AAD analytically computes the limit of the finite difference when the bump size goes to 0, hence also producing a sensitivity of 0. With finite differences using a larger bump, the resulting risk sensitivity depends on the proportion of paths that cross the money when the bump is applied, and this is a highly unstable number.

The risk of transactions with exotic discontinuities, such as barrier options, suffers from similar problems in the context of simulations. In this case, the estimate from N simulations is $\hat{V} = \frac{1}{N} \sum_{j=0}^{N-1} f\left(\vec{X}_j^a\right)$ where $\vec{X}_j^a$ is the scenario number j—the vector of all the market variables sampled by the model for all event dates from their joint risk-neutral distribution parameterized by a on simulation j—and f is the payoff as a function of the scenario. $\vec{X}_j^a$ is produced from a uniform sample of the hypercube $(0,1)^D$ (where D is the dimension of the Monte-Carlo, the number of random numbers for first path) with the application of a (generally continuous) function ϕ_a.[1] f is the function that computes the payoff of the transaction out of the scenario. In the context of a scripting language, f is the evaluation (against a scenario) of the script, and it is carried out by the evaluator visiting the expression trees resulting from the script. In the end, $\hat{V} = \frac{1}{N} \sum_{j=0}^{N-1} f\left[\phi_a\left(\vec{u}_j\right)\right]$ and it is clear that $\hat{V}$ is not differentiable in a when f is discontinuous.

[1]Classical Monte-Carlo simulations turn the sample $\vec{u}_j \in (0,1)^D$ into D independent Gaussian variables by the application of the inverse normal distribution to the components. Then it applies a transformation to produce correlated variables in accordance to model parameters, runs a discretization scheme, generally Euler's, also in accordance with model parameters, to produce the paths for state variables of the model, and finally computes the fixings for the simulated market variables $\vec{X}_j \in (0,1)^{D^*}$ as a function of the state variables and possibly the model parameters. We call ϕ_a the combined process that produces $\vec{X}_j \in (0,1)^{D^*}$ out of $\vec{u}_j \in (0,1)^D$ and depends on the model parameters a, generally in a way that is continuous in a.

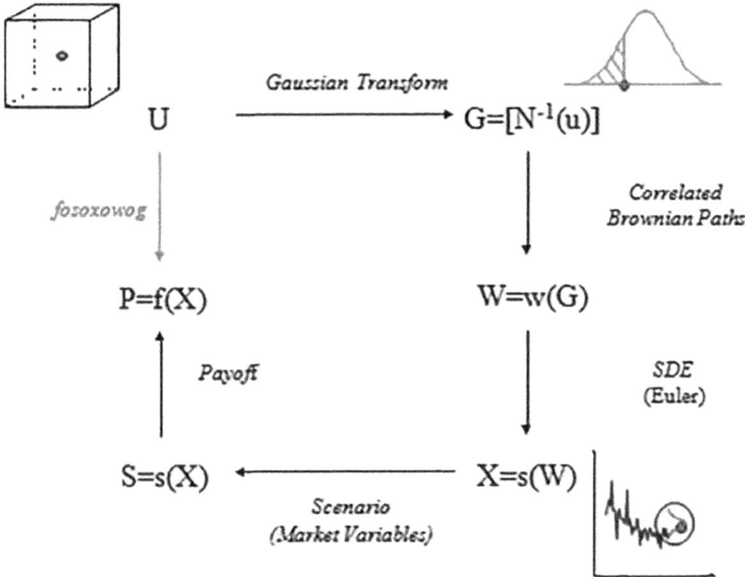

More sophisticated techniques may be applied, but they always boil down to some transformation ϕ_a from a.

16.2 TWO APPROACHES TOWARD A SOLUTION

The problem with Monte-Carlo and discontinuities is well known and the financial community has been dealing with it for decades. Two families of solutions have emerged. One works on the distribution φ_a of the scenario. The other focuses on the payoff f that is applied to the generated scenarios.

The first solution differentiates the general formula for the theoretical value of a transaction (possibly modulo discretization in time): $V = E^a\left[f(\vec{X})\right] = \int f\left(\vec{X}\right)\varphi_a\left(\vec{X}\right)d\vec{X}$ where φ_a is the joint (risk-neutral) distribution, parameterized by a, of all the components of the scenario: all the market variables on all the event dates. Therefore

$$\frac{\partial V}{\partial a} = \int f\left(\vec{X}\right)\frac{\partial \varphi_a\left(\vec{X}\right)}{\partial a}d\vec{X}$$

$$= \int f\left(\vec{X}\right)\frac{\partial \log \varphi_a\left(\vec{X}\right)}{\partial a}\varphi_a\left(\vec{X}\right)d\vec{X} = E^a\left[f\left(\vec{X}\right)\frac{\partial \log \varphi_a\left(\vec{X}\right)}{\partial a}\right]$$

In other terms, $\frac{\partial V}{\partial a} = E^a\left[g\left(\vec{X}\right)\right]$ where $g\left(\vec{X}\right) = f\left(\vec{X}\right)\frac{\partial \log \varphi_a\left(\vec{X}\right)}{\partial a}$. To compute a sensitivity we calculate the value of another payoff in the same model. The new payoff is the original payoff multiplied by the derivative of the log-likelihood of the scenario. This is like pricing a (somewhat twisted) path-dependent derivative transaction, and it can be done with standard Monte-Carlo simulations. Crucially, we end up differentiating the (generally differentiable) joint distribution instead of a non-differentiable payoff. Unfortunately, the practical implementation of this highly elegant solution is generally not convenient or practical. The path-wise calculation of $\frac{\partial \log \varphi_a\left(\vec{X}_j^a\right)}{\partial a}$ with Malliavin's calculus may be intractable or unstable depending on the model. We refer to Andreasen's Malliavin summary of 2007 [2] for details. Importantly, how this calculation is conducted depends on the model and its implementation, and it must be repeated for all models. For these reasons, the industry generally settled for the much simpler solution of approximating the payoff f with a "close enough" continuous function $\hat{f}$. The value is estimated with $V = E^a\left[f\left(\vec{X}\right)\right] \approx E^a\left[\hat{f}\left(\vec{X}\right)\right]$, which effectively removes discontinuities and resolves the problem. This solution performs remarkably well in practice, despite its simplicity, and is implemented in a model agnostic manner, hence its success.

We shall not discuss the Malliavin solution any further and focus on smoothing.

16.3 SMOOTHING FOR DIGITALS AND BARRIERS

Since discontinuities in European profiles are digital options, it suffices to approximate digital payoffs with continuous functions. Classically, digitals are approximated with tight call spreads. The following chat shows how a digital call (blue line) can be approximated by a call spread (red line) that has roughly the same value (subject to the spread size) and a continuous profile. The error in the approximation is related to the spread size and the profiles coincide at the limit as the spread size decreases to 0. Obviously, the spread size is a trade-off between the bias (on the value) introduced by the approximation when it is too large and the instability (on the risk) when it is too tight, causing the profile to revert toward discontinuity.

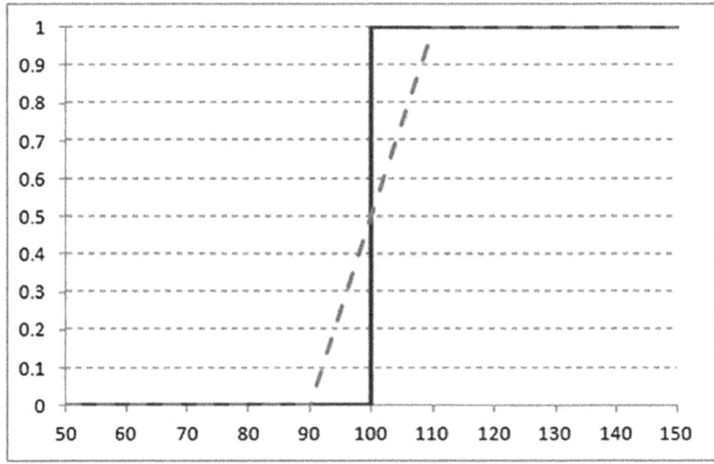

Call spread smoothing is very effective in practice in stabilizing risk sensitivities, both with finite differences and adjoint propagation. When the smoothing factor (or spread size) is well chosen (20 on the chart for clarity, in practice of order 1 in this case), the impact on the price is immaterial while risk sensitivities are stabilized. This is to our knowledge an industry-wide standard for smoothing risk sensitivities with Monte-Carlo simulations.

Let us discuss further this industry standard "call-spread" smoothing technique, and in particular provide three distinct interpretations that will be very useful in what follows.

Fuzzy logic With conventional—or sharp or crisp—logic, the digital either ends up in the money and pays \$1, or ends up out of the money and pays 0, hence the discontinuity. Reasoning with the so-called fuzzy logic instead, we regard a fixing above 110 as decidedly in the money, a fixing below 90 as decidedly out of the money, and any fixing in between as partly in the money, a bit like Schrodinger's cat who is part dead in its capsule, with a "degree of moneyness" that is interpolated between 0% at 90 and 100% at 110. The call spread smoothing corresponds to a transaction that pays \$1 over the fraction of the notional that ends up in the money. We investigate fuzzy logic further in subsequent chapters, where it leads to an efficient, elegant automatic smoothing algorithm.

Finite difference A digital profile is the derivative (in the sense of distributions) of European call profiles with respect to strikes: $D_K(S) \equiv 1_{\{S>K\}} = -\frac{\partial}{\partial K}(S-K)^+ = -\frac{\partial}{\partial K}C(S,K)$. The call spread smoothing is nothing else than a finite difference approximation of the derivative. At the limit, the derivative is discontinuous, but before the limit, all finite difference approximations are continuous, hence the validity and efficacy of the technique.

Spreading of notional We invert the previous equation into $C(S,K-\varepsilon) - C(S,K+\varepsilon) = \int_{K-\varepsilon}^{K+\varepsilon} D_k(S)dk$. A call spread is a collection of digitals with strikes evenly distributed within the spread. Hence, to apply call spread smoothing also corresponds to booking a collection of digitals of different strikes instead of one digital with a given strike. The notional of the original transaction is evenly spread over digitals stroke within the spread, hence the smoothing.

The industry generally applies the same approach to discontinuous exotics, in particular transactions with barriers. A transaction with an up-and-out barrier, for instance, has a schedule of cash-flows that stops as soon as the spot fixes above a level B, causing a discontinuity in the profile. To resolve the problem, such as for digitals, we spread the notional of the transaction across barriers in a region around B. On all barrier-fixing dates, we potentially knock out only a fraction of the notional interpolated between 0 on or below $B-\varepsilon$ and 100% on or above $B+\varepsilon$. We carry on with the schedule of cash-flows with the remaining notional, which in turn is susceptible to be knocked-out in part or in full on the next barrier fixing. Like Schrodinger's cat, our transaction is partly alive or *alive to a degree.* The natural interpretation for this "degree of aliveness," however, is *not* the probability of being alive, but the fraction of the notional that remains. This extension of call spread smoothing to barriers is an industry standard that is sometimes referred to as "smooth barriers."

16.4 SMOOTHING FOR SCRIPTED TRANSACTIONS

The smoothing of digitals and barriers is widely used in the industry. This is a standard that a professional scripting language must incorporate and support. With scripted products, it is always the conditional blocks starting with "if" statements that produce discontinuities. According to the condition, different statements are executed and different, potentially discontinuous values, are assigned to variables and paid by products. Scripts typically involve frequent conditional statements, and each one is the source of a potential discontinuity that breaks risk sensitivities with simulations. As a consequence, and while scripting has been a major progress for valuation, the risk management of scripted transactions has often been unstable and untrustworthy unless some form of smoothing is performed. When a scripted transaction with conditional statements is booked in view of risk management, the script has typically been manually modified to incorporate smoothing. However, this manual approach to smoothing discussed next is somewhat unsatisfactory and we subsequently investigate automatic solutions.

CHAPTER 17

Support for Smoothing

We first discuss support for manual smoothing in the scripting syntax. This is easily implemented with an additional function that we may call "functional if," or *fIf*, similar to Excel's *if* but with a smoothing factor that can be used to replace *if* statements. The signature of the function is:

```
fIf( x, value if x positive, value if x negative, smoothing factor)
```

It is formally defined as:

$$fIf(x, a, b, \varepsilon) = b + \frac{a-b}{\varepsilon}\max\left(0, \min\left(\varepsilon, x + \frac{\varepsilon}{2}\right)\right)$$

and drawn here:

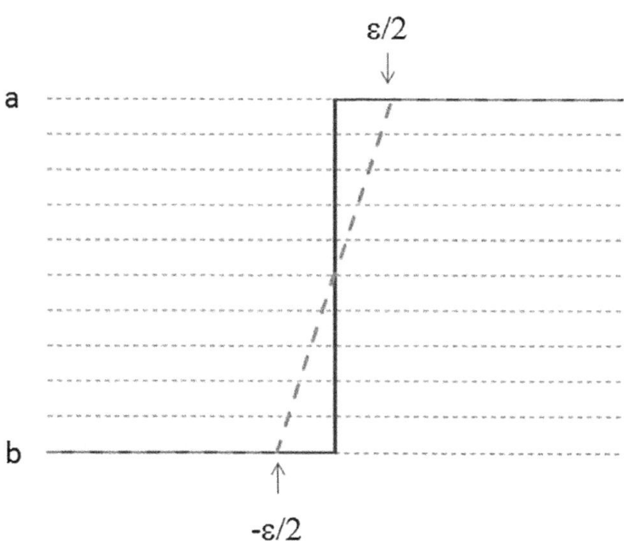

This function can be implemented in the scripting language in the same way as any other function: *exp*, *log*, *sqrt*, and so on. We explained how to extend the scripting syntax with additional functions in part I, and this one is a particularly lightweight development. Once implemented in the syntax, the function can be used in scripts to replace *if* statements and apply a call spread smoothing to their conditions. For a digital (with a smoothing factor of 1) we would have the smoothed script:

```
01Jun2021    dig pays fIf( Spot() - 100, 1, 0, 1)
```

This effectively removes the instabilities in risk sensitivities. Discontinuities produced by *if* statements are remedied by *fIf*. The following charts show prices, deltas, and vegas in Black-Scholes with 0 rates and 20% volatility for the initial script and the smoothed script. The seed is reset between simulations so as to offer a visual glimpse at the Monte-Carlo error. The improvement in sensitivities is immediately visible while there is no visible impact on the value.

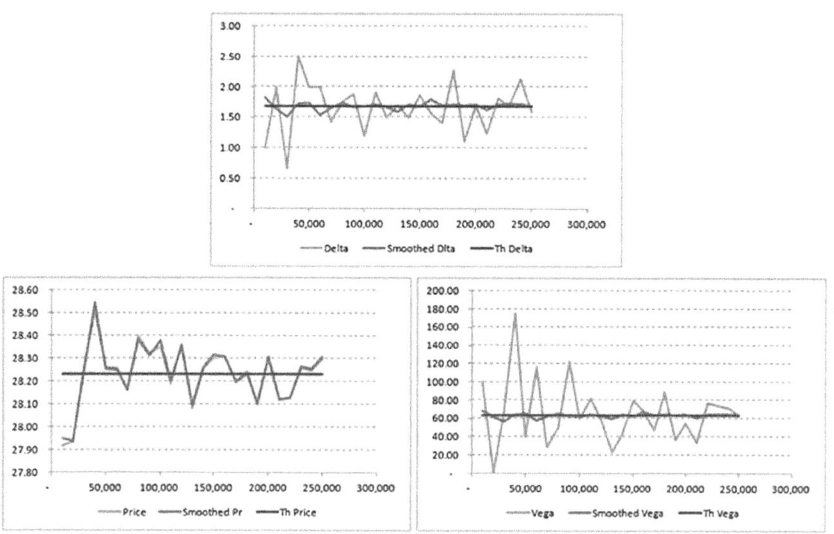

The *fIf* syntax also applies to the smooth barrier approximation. An original script for a call with an up-and-out (discretely monitored) barrier like:

01Jun2020	vAlive = 1
start: 01Jul2020 end: 01Jun2021 freq: 1m fixing: end	if spot() > BARRIER then vAlive = 0 endIf
01Jun2021	opt pays vAlive * CALL(spot(), STRIKE)

may be rewritten with functional conditions:

01Jun2020	vAlive = 1
start: 01Jul2020 end: 01Jun2021 freq: 1m fixing: end	vAlive = vAlive * fIf(spot() - BARRIER, 0, 1, EPS)
01Jun2021	opt pays vAlive * CALL(spot(), STRIKE)

This also effectively removes risk instabilities with a minimal impact on the price (with a carefully chosen EPS). The following charts show the price, delta, and vega in Black-Scholes (spot = 100, rates = 0, volatility = 20%) for a 110 1Y call 120 RKO, weekly monitored (with the barrier shifted down to 118.36 to compensate for weekly monitoring in accordance to the 0.6*std* rule). We use a smoothing factor of 1.

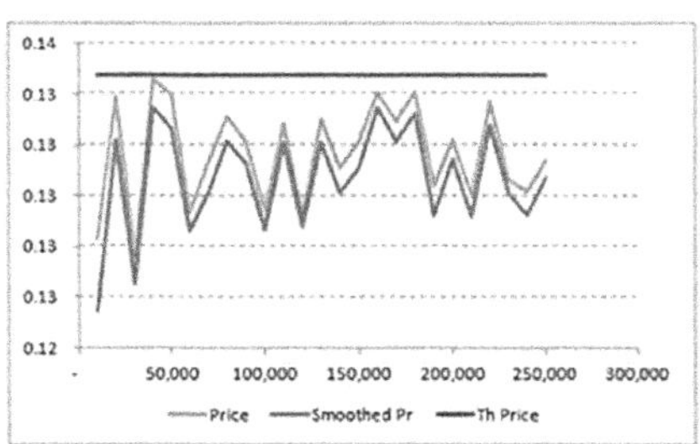

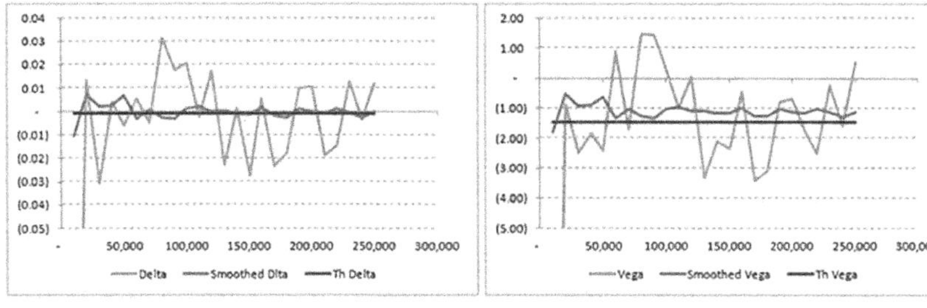

Finally, our simplified autocallable:

01Jun2020	vRef=100 vAlive=1

01Jun2021	if spot() > vRef then
	prd=110
	vAlive=0
	endIf

01Jun2022	if spot() > vRef then
	if vAlive=1 then prd=120 endIf
	vAlive=0
	endIf

01Jun2023	if vAlive=1 then
	if spot() > vRef then prd=130 else prd=50 endIf
	endIf

may be smoothed with (smoothing factor is 1 again):

01Jun2020	vRef=100 vAlive=1
01Jun2021	prd pays fIf(spot() - vRef, 110, 0, 1) vAlive = fIf(spot() - vRef, 0, 1, 1)
01Jun2022	prd pays vAlive * fIf(spot() - vRef, 120, 0, 1) vAlive = vAlive * fIf(spot() - vRef, 0, 1, 1)
01Jun2023	prd pays vAlive * fIf(spot() - vRef, 130, 50, 1)

Note once again the interpretation in terms of fuzzy logic. The transaction is partly alive and delivers an interpolation between the schedule of "if alive" coupons and the schedule of "if dead" coupons.

And it works, as seen on these charts:

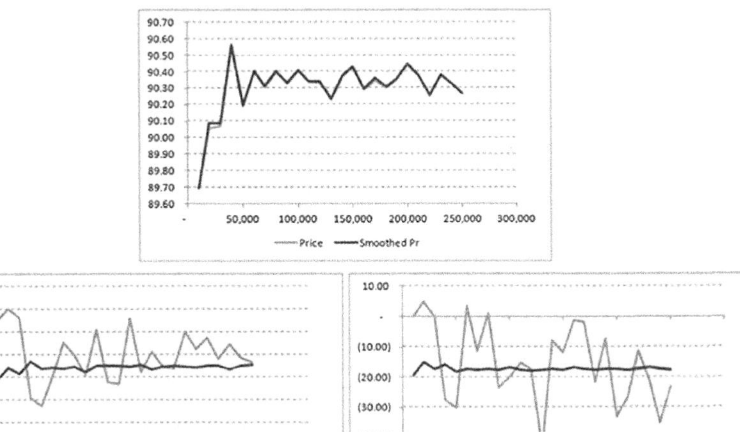

But it also clearly demonstrates a limit to this manual approach. The smoothed script is awkward, unnatural, and prone to error. It certainly does not fulfill our goal of a raw description of cash flows in a natural, accessible syntax. Yet, virtually every institution that uses a scripting system for derivatives smooths scripts manually. Practitioners typically write natural scripts for pricing and turn them into functional scripts for risk management. This process is time-consuming and prone to error. Our simplified autocallable is an extremely simple transaction with hardly any exotic features. With more complex transactions, manual smoothing can turn into an operational nightmare. What we really need is an automated algorithm for smoothing scripts.

CHAPTER 18

An Automated Smoothing Algorithm

In the context of scripting, discontinuities always come from *if* statements.[1] We assume for now that we know what smoothing factor EPS we want to apply, and that the functional if, *fIf*, is implemented and available. Without loss of generality, we also assume that all elementary conditions are represented as

expression > 0

or

expression > = 0

or

expression = 0

This is a trivial transformation of general elementary conditions *expr*1[<][>][=]*expr*2 that can be implemented in a dedicated visitor at processing time, or even at parsing time with small modifications in the *parseCondElem*() function developed in appendix A.

Finally, and more importantly, we assume for now that the *condition domain*, that is, the value domain of *expression* in *expression* > [=]0, is a continuous interval that includes 0. This is the case for conditions of the type

[1]Unless we implemented in the language functions that are themselves discontinuous. In that case, smoothing these functions is trivial. Here we discuss the less obvious matter of smoothing a general script, where every *if* statement produces a potential discontinuity.

spot() > 100 but not *vAlive* = 1, where the LHS expression *vAlive* has discrete domain $\{0, 1\}$. We deal with discrete domains in chapter 20. For now, we note that in the case of continuous domains, equalities are meaningless and inequalities are always strict. Hence, the only elementary conditions of interest here are strict inequalities of the form *expression* > 0.

18.1 BASIC ALGORITHM

We want to design an algorithm to apply smoothing automatically, that is, turn a natural script, one with potentially discontinuous conditions, into a smoothed script, where all conditions are functional. For example (this script is purposely structured for the exercise and obviously meaningless otherwise):

Date1	... (no conditional statements) ...

Date2	if spot() - x > 0 then
	y = x
	z = y
	t pays z
	else
	z = 0
	r = 1
	endIf

Date3	... (no conditional statements) ...

We know that no change is necessary for unconditional statements. We transform only conditional blocks "if... then... [else]... endif," including conditional statements and nested conditional blocks if any. In our toy script, only the event for Date2 needs work. Intuitively, it could be smoothed with the somewhat simplistic (but flawed) algorithm (where EPS is our smoothing factor):

1. Replace elementary conditions by functional equivalents to compute the degree of truth *degTrue* of the condition. Denote

$$degFalse = 1 - degTrue$$

2. In all *iftrue* statements (those to be executed if the condition is true):
   (a) Replace assignments "X = RHS" by "X = degTrue*RHS + degFalse*X."
   (b) Replace payments "X pays RHS" by "X pays degTrue*RHS."

3. Apply the same procedure to the *iffalse* statements, weighting with *degFalse*.

   The result would look like:

```
      degTrue = fIf( spot() - x, 1, 0, EPS)
      degFalse = 1 - degTrue

      y = degTrue * x + degFalse * y
Date2 z = degTrue * y + degFalse * z
      t pays degTrue * z
      z = degFalse * 0 + degTrue * z
      r = degFalse * 1 + degTrue * r
```

This script does not produce the correct results. For instance, the final value of y in this case is correctly $degTrue * x0 + degFalse * y0$, but z ends up at $degTrue^3 * x0 + degTrue^2 * degFalse * y0 + degTrue * degFalse * z0$, when we should expect just $degTrue * x0$.

The dependencies between the variables in the conditional statements, and the assignments to the same variables in the *iftrue* and *iffalse* statements, interfere with our simplistic logic and cause the algorithm to fail.

We fix the algorithm by recording the values $X0$ of all the variables before the execution of the conditional block, and use these stored, untouched values in statements *a la* "X = degTrue*RHS + degFalse*X0." The algorithm becomes:

1. Replace elementary conditions by functional equivalents to compute the degree of truth *degTrue* of the condition. Denote

$$degFalse = 1 - degTrue$$

2. Store the initial values (X0) of the variables before the evaluation of the conditional block.
3. Evaluate all *iftrue* statements.
4. Reset variables to (X0).
5. Evaluate all *iffalse* statements, if any.
6. We call $(X2)$ the values of the variables after the two previous steps. If there were no *iffalse* statements, $(X2) = (X0)$. Otherwise, $(X2)$ holds the final values after evaluation of the *iffalse* statements, starting with state (X0).
7. The final values of the variables is:

$$(X) = degTrue * (X1) + degFalse * (X2)$$

Now our transformed script looks like:

Date2	Correct smoothing degTrue = fIf(spot() - x, 1, 0, EPS) degFalse = 1 - degTrue Store initial state x0 = x y0 = y z0 = z t0 = t r0 = r Apply *iftrue* statements y = x z = y t pays z Store *iftrue* results x1 = x y1 = y z1 = z t1 = t r1 = r Reset initial state x = x0 y = y0 z = z0 t = t0 r = r0 Apply *iffalse* statements z = 0 r = 1 Apply fuzzy logic x = degTrue * x1 + degFalse * x y = degTrue * y1 + degFalse * y z = degTrue * z1 + degFalse * z t = degTrue * t1 + degFalse * t r = degTrue * r1 + degFalse * r

We can check that on exiting that block of instructions, all variables have their correct expected values. We can also check that the algorithm correctly smooths the digital and barrier scripts, producing the exact same scripts we wrote manually in chapter 17. The autocallable script, however, cannot be smoothed by this algorithm alone because it has equality conditions on the *boolean* variable *vAlive*. This condition's domain is not continuous, hence the application of the call spread in the *fIf* function does not make sense. For this script, we need our final algorithm, one that can handle discrete condition domains. That final algorithm is presented in chapter 20.

18.2 NESTED AND COMBINED CONDITIONS

We may also verify that our algorithm correctly deals with nested conditions. For instance, with an inner conditional block among the *iftrue* statements of the outer block, the inner block's *iftrue* statements are weighted by the product of the degrees of truth (*DT*)s of the outer and inner conditions. Our algorithm implicitly processes nested conditions by multiplying the *DT* of the inner condition by the *DT* of the outer condition. And since nested conditions correspond to an *and* operator, this means that the algorithm implicitly applies:

```
(1) dt( A and B) = dt ( A) * dt ( B)
```

For consistency, we should apply the same for conditions explicitly combined with *and*.

We also used *degFalse* = 1 − *degTrue*, so we applied:

```
(2) dt ( not A) = 1 - dt ( A)
```

We see that *DT*s combine like probabilities, so for consistency we would also use:

```
(2) ( A or B) = dt ( A) + dt ( B) - dt ( A) * dt ( B)
```

Formulas (1), (2), and (3) provide the computation of the *DT* of combined conditions as a function of the *DT*s of their elementary conditions, which completes our algorithm.

18.3 AFFECTED VARIABLES

We now have a fully working smoothing algorithm for continuous conditions, but it may be terribly inefficient. We are storing intermediate values of all variables involved in the script, twice, for every conditional block. Scripts for xVA or other portfolio-based risk calculations may have hundreds of thousands of variables. Should we copy them all (twice!) for the evaluation of every conditional block, where some may modify the value of one variable only?

The only variables that should be recorded and restored as part of our algorithm are the variables that are *affected* in the condition block, meaning variables that are assigned or paid into in the *iftrue* or *iffalse* statements of the block, including nested blocks. For instance, in our earlier toy example, x is not affected and does not need to be recorded or restored.

Hence, in order to optimize our algorithm, we must identify the affected variables for every conditional block. This can be implemented without difficulty in a dedicated pre-processor. Then, a small modification is sufficient to restore our algorithm's efficacy:

1. Compute the degree of truth *degTrue* of the condition. Denote
$$degFalse = 1 - degTrue$$
2. Store the initial values $(X0)$ of the *affected* variables before evaluation of the conditional block.
3. Evaluate all *iftrue* statements.
4. Store the final values $(X1)$ of the affected variables after evaluation of the *iftrue* block.
5. Reset affected variables to $(X0)$.
6. Evaluate all *iffalse* statements (if any).
7. We call $(X2)$ the values of the *affected* variables after the two previous steps. Compute the final values of the affected variables after evaluation of the condition block with:
$$(X) = degTrue * (X1) + degFalse * (X2)$$

Evidently, unaffected variables don't need any modification.

18.4 FURTHER OPTIMIZATION

Finally, we note an important optimization that reduces the performance penalty of smoothing to virtually nothing in most cases. The evaluation of the modified script may take substantially longer than the original, because both the *iftrue* and *iffalse* blocks of statements must be evaluated unconditionally, in all scenarios, where in the original script, we evaluated only one of them (or none). In practice, however, because we use a call spread smoothing, *degTrue* is 0 or 1 in most scenarios, so either the *iftrue* or the *iffalse* statements are uselessly executed only to be applied a weight of 0 in the end. It is only in the low probability event where the expression in the condition lands between $-EPS/2$ and $EPS/2$ that both blocks must be evaluated. With this in mind, we optimize the algorithm to produce a smoothed script that evaluates virtually as fast as the original script and therefore offers smoothing for free:

1. Compute the degree of truth *degTrue* of the condition. Denote
$$degFalse = 1 - degTrue$$

2. If *degTrue* = 1, evaluate all *iftrue* statements and exit.
3. If *degTrue* = 0, evaluate all *iffalse* statements and exit.
4. Else (if $0 < degTrue < 1$):
   (a) Store the initial values ($X0$) of the affected variables.
   (b) Evaluate all *iftrue* statements.
   (c) Store the final values ($X1$) of the affected variables.
   (d) Reset affected variables to ($X0$).
   (e) Evaluate all *iffalse* statements (if any).
   (f) We call ($X2$) the values of the affected variables after the two previous steps. Compute the final values of the affected variables after evaluation of the condition block with:

$$(X) = degTrue * (X1) + degFalse * (X2)$$

We should also optimize the pre-processing of the affected variables. That pre-processor should not only identify the affected variables in conditional blocks but also store their indices on the corresponding *if* node. The same pre-processor, or another one, should also pre-allocate space for the storage of the affected variables so no memory allocation needs to happen at simulation time. The implementation is not trivial because it must deal with potential nested conditional blocks.

The entire source code is accessible in our online repository. The pre-processor is called *IfProcessor* and the code is in scriptingIfProc.h. The run-time algorithm is part of the derived *FuzzyEvaluator* in scriptingFuzzy Eval.h. All the optimizations discussed are part of the code.

This is the final version of the automated smoothing algorithm. We call it the *Fuzzy Condition Evaluator or FCE* for reasons that will be made apparent soon. It works with all conditions, including nested and combined, provided that all the elementary conditions have continuous domain, which is the value domain of *expression* in *expression* > 0, and a continuous interval that includes 0. We extend it to general domains in chapter 20. For now, we note that in the case of digitals and barriers, this algorithm exactly replicates classical smoothing. For other transactions with discontinuities, the algorithm also replicates the smoothing traders would typically (and correctly) apply manually. In other words, fuzzy evaluation of conditional blocks automatically implements smoothing in the same way traders do manually and effectively removes risk instabilities due to discontinuities.

CHAPTER 19
Fuzzy Logic

Our simple algorithm (FCE) effectively turns a natural script into a smoothed one, provided all conditions are continuous. With a well-chosen smoothing factor, the impact on the price is immaterial, and risk sensitivities converge nicely with Monte-Carlo. Performance is virtually unchanged.

How are we going to implement the FCE? Are we really going to traverse scripts and turn them into equivalent smooth ones like we did in chapter 18? The new script would be harder to read, similar to anything that is written by computers. In practice, we would have to keep the original script and store it along the modified one. Or, the modification could occur on demand, when risk sensitivities are being computed, but this would produce an overhead. To modify scripts for computational purposes violates the modularity of our code and the principle of a loose coupling between the cash flows and their evaluation. We would much rather apply smoothing *without modification of the script*. Here, we show how the FCE may be implemented without changing the script, using a modification in the evaluator instead.

Reviewing the steps involved in the FCE, we can see them as a different *evaluation* of the same script rather than a modification of the script. Further, it is only the conditional blocks that are evaluated in a different way, the rest of the script being evaluated identically. Hence, the script does not need to change provided we modify the evaluator to implement a different *logic* for the evaluation of conditional blocks. In practice, this means that we *derive* the evaluator and *override* its methods related to visits to *if* nodes.

We developed visits to *if* nodes in our base evaluator to perform the following steps:

1. Evaluate the condition.
2. If the condition is true, evaluate the *iftrue* statements.
3. Otherwise, evaluate the *iffalse* statements if any.

In our derived evaluator, we override the visit to the *if* nodes so they implement the steps of the FCE instead. We also override the visits to conditions and condition combinators *and*, *or*, and *not* to compute degrees of truth (DTs) and evaluate conditions to a number between 0 and 1, instead of *true* or *false*. Then we use the derived evaluator to value the script in every scenario, effectively implementing FCE without any modification to the script.

It is clear that what changes is the *logic* at play in the evaluation of conditions and conditional blocks. Interestingly, such overriding exactly consists of the application of *fuzzy logic* in place of conventional logic. Hence, our derived evaluator that smooths scripts at evaluation time is indeed a *fuzzy evaluator*, an evaluator that applies fuzzy logic to conditions and conditional statements. This should not come as a complete surprise, since we had already noted the interpretation in terms of fuzzy logic of the classical smoothing of digitals and barriers.

Fuzzy logic is an extension of conventional logic that was invented in the 1960s by Lotfi Zadeh, who spent the following 30 years further developing and promoting it. His major publications are compiled in [29]. The initial goal was an attempt to model expert opinions for the development of the first expert systems. Fuzzy logic extends conventional logic, where conditions are either *true* or *false*, with the notion that a condition may be true *to some extent*. The extent to which a condition is true is a number between 0 and 1 called its *degree of truth* (we write DT in short). The DT of some condition $expression > 0$ is a function of the expression: $DT = dt(expression > 0) = f(expression)$. We choose to calculate DTs with call spreads for efficiency and consistency with industry standards. Alternatively, we could use any continuous non-decreasing function valued into (0,1). It is desirable to parameterize the function to control how fast it increases from 0 to 1 as the expression increases from very negative values resulting in a DT of 0 to very positive values resulting in a DT of 1. In the case of call spreads, the parameter is the size of the spread EPS. Other examples of valid, relevant forms for DT include cumulative-centered Gaussian distributions parameterized by a standard deviation s or a logistic function parameterized by its steepness k. Such choices offer additional smoothness, but they deviate from market practice and more importantly produce a significant performance overhead. Hence, we stick with a call spread specification, but point out that the FCE works with any sensible specification. In practice, the computation of the DTs is implemented in the overridden visitor to the > node (elementary conditions) in the derived evaluator.

Fuzzy logic also provides computations for the DT of combined conditions. We choose *probabilistic style* computations

```
dt( A and B) = dt( A) * dt( B)
dt( A or B) = dt( A) + dt( B) − dt( A) * dt( B)
dt( not A) = 1 − dt( A)
```

for consistency with how we deal with nested conditions. Another popular choice among fuzzy logic practitioners is the *minMax* form:

```
dt( A and B) = min( dt( A), dt( B))
dt( A or B) = max( dt( A), dt( B))
dt( not A) = 1 – dt( A)
```

In practice, the computation of the combined DT is implemented in the overridden visitor to the *and*, *or*, and *not* nodes in the derived evaluator.

Finally, fuzzy logic propagates fuzziness to affected variables. Consider the following statement similar to the autocallable script at maturity:

```
Ii spot() > 100 then vDigPays = 130 else vDigPays = 50 endIf
```

With sharp logic, *vDigPays* is a binary variable that is evaluated to either 50 or 130 depending on the condition. Its (discrete) value domain is {50,130}. But when this statement is evaluated in light of fuzzy logic, *vDigPays* is interpolated between 50 and 130 according to the DT of the condition. Implicitly, the statement becomes:

```
vDigPays = dt( spot() - 100) * 130 + [1 – dt( spot() – 100)] * 50
```

vDigPays becomes a continuous variable valued in the real interval (50,130). Fuzzy logic applied to the condition propagated into *vDigPays*. We say that *vDigPays* has been *fuzzed*. This matters, because a later statement in the script may test *vDigPays*:

```
If vDigPays = 50 then prd pays [payoff1] else prd pays [payoff2] endIf
```

The condition *vDigPays* = 50 no longer makes sense after *vDigPays* has been made continuous. Furthermore, two tests, such as *If vDigPays* > 100 and *If vDigPays* > 120, which were equivalent in light of conditional logic, are no longer equivalent. Chapter 20 deals with domains and the propagation of fuzziness and resolves most of the difficulties.

For now we take note of the three characteristics of fuzzy logic in play in our evaluation of scripts:

1. The evaluation of the DT of elementary conditions *expression* > 0 as continuous non-decreasing functions valued in $(0, 1)$, for example, call spreads with size EPS.

2. The computation of the DT of combined or nested conditions from the DTs of their elementary conditions, for example, with probabilistic formulas.
3. The propagation of fuzziness into the variables that are affected in conditional statements. These variables are implicitly interpolated (or *fuzzed*) between their values *iftrue* and their values *iffalse*, according to the DT of the condition.

To evaluate the conditional blocks in our scripts with fuzzy logic, or to modify scripts with the FCE, obviously produces the same results. It is remarkable that the smoothing of financial payoffs, as it has been conducted by the industry for decades, exactly coincides with the evaluation of conditions and conditional statements under fuzzy logic.

More importantly, this realization leads us toward a general, simple, efficient, and elegant implementation of the algorithm by derivation of the evaluator, and the implementation of overridden visitors to conditional blocks using fuzzy logic. Concretely, we derive the Evaluator class into a "FuzzyEvaluator" that overrides the following visitors:

Elementary conditions We remind that all inequalities are transformed at parsing or processing time into equivalent conditions *expression* > 0. Hence, we only derive the visitor to the > node. Instead of evaluating elementaries to *true* or *false* and pushing the result onto the boolean stack, the derived evaluator evaluates $dt = fIf(expression, 1, 0, EPS)$[1] and pushes the result onto a dedicated stack of numbers between 0 and 1 (the *dtstack*).

Combinators The derived evaluator computes the DT of combined conditions from the DTs of the elementary conditions involved, for example, with probabilistic formulas, implemented with overridden visitors to combinators *and*, *or*, and *not*.

Conditional blocks The derived evaluator overrides visitors to *if* nodes to implement fuzzy logic with the FCE algorithm.

Our entry-level evaluation function takes a parameter so that either sharp or fuzzy logic is applied. When valued with sharp logic we proceed with the base evaluator. With fuzzy logic, we use an instance of the derived *FuzzyEvaluator* instead. The script remains unchanged. We can evaluate it with sharp and fuzzy logic and compare the stability of sensitivities, the values, and computation time. In the risk management system, we can

[1]Since the value if positive is always 1 and value if negative is always 0, we actually implement an optimized version of *fIf*.

systematically use sharp logic for valuation and fuzzy logic for risks on the same script that is booked and processed only once.

The source code for the fuzzy evaluator is in scripting FuzzyEval.h. The code for the processor that conducts the necessary precomputations and pre-allocations is in scriptingIfProc.h. Chapter 20 deals with the more difficult processing of domains. In fact, we use two different processors in a sequence for domains. The sources for both are in scriptingDmainProc.h and scriptingConstCondProc.h.

CHAPTER 20

Condition Domains

We designed in the previous chapters a general smoothing algorithm (FCE) that works as long as all conditions have continuous domains. This is not good enough. We cannot smooth our simplified autocallable script because it contains conditions on discretely valued expressions. We now address this problem and complete our algorithm.

The FCE starts with the computation of the degrees of truth (DTs) of all conditions. The DTs of elementary conditions are given by *fIf*. That formula makes sense only for continuous conditions. Everything that follows—the computation of combined conditions with probabilistic style combination formulas, and the evaluation of conditional blocks, including the nested ones, by the FCE—is correct as long as the elementary DTs are correct. So far, we only know the formulas for the DTs of elementary conditions. Now, we investigate the DTs of conditions whatever their domain.

First, we discuss the evaluation of the DT of conditions other than continuous, provided their domain is known. Then, we discuss how we practically identify the condition domains with dedicated pre-processors.

20.1 FUZZY EVALUATION OF DISCRETE CONDITIONS

20.1.1 Condition domains

Consider a condition of the type $e > 0$, $e >= 0$, or $e = 0$ where e is some expression that evaluates to a number. These three forms cover all elementary conditions provided we perform necessary transformations at parsing or pre-processing time.

We call *domain of the condition*, or *domain of the expression*, and denote *dom*(e) the set of all possible values for *expression* when the script is analyzed in light of conventional logic.

When *dom(e)* is the real space, or at least includes a continuous interval that contains 0, we call the condition domain *continuous*. In this case, equalities don't make sense and inequalities are necessarily strict. The fuzzy evaluator as designed in chapter 19 works with all continuous conditions.

We now explore how we compute the DT of other types of conditions, starting with the simplest kind, the *constant* conditions, where *dom(e)* is a single number, and then making our way toward more general ones.

20.1.2 Constant conditions

When *dom(e)* is the singleton $\{c\}$, we call the condition *constant*. Equalities make sense and inequalities are not necessarily strict. Our fuzzy evaluator fails to correctly evaluate constant conditions. Consider the following example:

Date1	vCst = 1
Date2	if vCst > = 1 then then x = 1 else x = 0 endIf

This script may seem irrelevant at first sight, but in practice constant tests are frequently used to quickly switch on and off particular cash-flows or features in a script. If we were to fire our fuzzy evaluator on $vCst >= 1$, it would (implicitly) evaluate a call spread around 0 in $vCst - 1$, which always evaluates to 0, hence, the call spread always evaluates to 0.5, resulting in a DT of 0.5 and a final result of $x = 0.5$. This is obviously wrong, the correct result trivially being $x = 1$.

The right thing to do with constant conditions is immediately evaluate them to true or false (DT = 1 or 0) and skip the evaluation altogether. When $dom(e) = \{c\}$, we know that $dt(e > [=]0) = 1$ if $c > [=]0$, 0 otherwise, independently of the scenario.

The correct fuzzy evaluation of the DT of elementary conditions depends on each condition's domain. When the domain is continuous, the FCE applies unmodified. When the domain is discrete, we need to proceed with a different calculation. When it is constant, in particular, the condition must be marked as *alwaystrue* or *alwaysfalse* so its evaluation reads its pre-calculated DT and skips the evaluation of the condition.

It is clear that in order to proceed, we must compute the domains of all the elementary conditions in the script, and in particular identify whether their domain is continuous or discrete, and whether the evaluation of the condition may be pre-calculated as *alwaystrue* or *alwaysfalse*. We will design a visitor that computes condition domains before simulations take place. For each elementary condition in the script, the visitor will write its domain on the condition's node and also flick its *alwaystrue* and *alwaysfalse* flags. Before we get there, let us continue our investigation of the fuzzy evaluation of conditions with discrete domains.

20.1.3 Boolean conditions

When $dom(e) = \{0, 1\}$, we call the condition *boolean*. In this case, like in the constant case, equalities make sense and inequalities may not be strict. Our fuzzy evaluator also fails as in the following example:

Date1	Continuous if spot() > 100 then vAlive = 1 else vAlive = 0 endIf
Date2	Boolean if vAlive > 0 then x = 1 else x = 0 endIf

Boolean conditions are very frequent in real-world scripts, and we cannot get away with an algorithm unable to deal with them properly. To apply the FCE in this case leads to (completely) wrong results. Our computations would (implicitly) apply a call spread of size ε_S (of order of magnitude 1) to $spot() - 100$ leading to

$$vAlive = 1_{\left\{100-\frac{\varepsilon_S}{2}<S<100+\frac{\varepsilon_S}{2}\right\}} \frac{S - 100 + \frac{\varepsilon_S}{2}}{\varepsilon_S} + 1_{\left\{S>100+\frac{\varepsilon_S}{2}\right\}}$$

Then $vAlive > 0$ would be evaluated with a call spread of size ε_V, finally resulting in

$$\begin{aligned} x &= 1_{\left\{-\frac{\varepsilon_V}{2}<vAlive<\frac{\varepsilon_V}{2}\right\}} \frac{vAlive+\frac{\varepsilon_V}{2}}{\varepsilon_V} + 1_{\left\{vAlive>\frac{\varepsilon_V}{2}\right\}} \\ &= 1_{\left\{100-\frac{\varepsilon_S}{2}<S<100-(1-\varepsilon_V)\frac{\varepsilon_S}{2}\right\}} \left(\frac{S-100+\frac{\varepsilon_S}{2}}{\varepsilon_S\varepsilon_V} + \frac{1}{2}\right) + 1_{\left\{S>100-(1-\varepsilon_V)\frac{\varepsilon_S}{2}\right\}} \end{aligned}$$

We draw here the final value of x as a function of the spot fixing for different choices of ε_V:

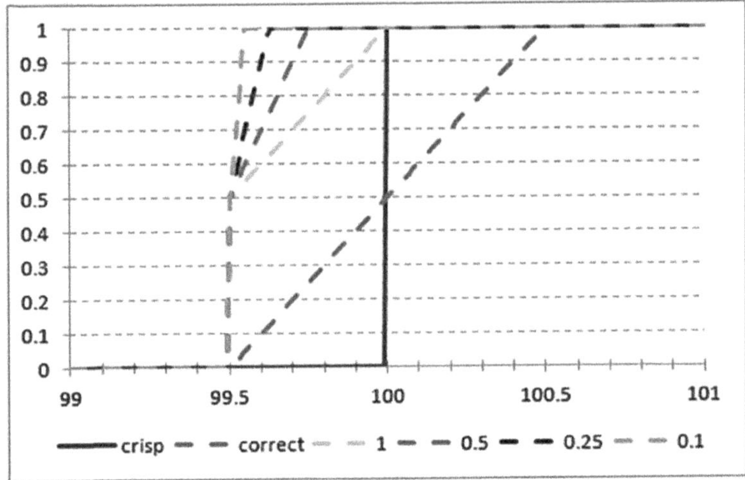

The bias is very clear. Not only are we severely overestimating x, it gets worse as we reduce ε_V toward 0. The bias is actually maximal on convergence.

We cannot apply the continuous domain formulas to boolean conditions. So what is the correct formula for $dt(vAlive > 0)$? In our script, x evaluates to $dt(vAlive > 0)$ and its correct value is evidently $dt(spot() > 100)$. Hence:

$$dt(vAlive > 0) = dt(spot() > 100) = vAlive$$

Remember, *vAlive* is *fuzzed* on date 1 and its fuzzed domain is $(0, 1)$ when $vAlive > 0$ is evaluated. When $vAlive = 0$, it is certainly not true that $vAlive > 0$ and the DT is 0. When $vAlive = 1$, it is certainly superior to 0, and the DT is 1. In between, the DT is interpolated between 0 on $vAlive = 0$ and 1 on $vAlive = 1$. Therefore, the DT of $vAlive > 0$ is *vAlive* itself.

The correct DT of a boolean condition $e > 0$ is

$$dt(e > 0) = e$$

As an important side note, reading the event 1 in light of sharp logic leads to the domain $\{0, 1\}$ for *vAlive*. But in light of fuzzy logic, *vAlive* is continuous with domain $(0, 1)$. Yet, our algorithm for continuous domains fails to correctly evaluate the condition on date 2. Even though fuzzy evaluation makes all domains continuous (except those of constant expressions), what matters is the domain derived from a *crisp reading* of the script. Here, the domain is $\{0, 1\}$, not $(0, 1)$. It is a discrete domain, hence the computation for continuous conditions fails and a different calculation applies where $dt(e > 0) = e$.

20.1.4 Binary conditions

More generally, when $dom(e) = \{-a, b\}$ in $e > 0$, we call the condition *binary*. We must have $-a <= 0$, $b >= 0$, and $a \neq b$, otherwise the condition is always true or always false, hence constant, not binary. The FCE fails to evaluate correctly the DT of binary conditions, as was clearly demonstrated for the boolean case. The correct evaluation of the DT in this case is by an elementary extension of the previous result:

$$dt(e > 0) = \frac{a + e}{a + b}$$

In the boolean script from the previous paragraph, we could have written "if vAlive > 0.5" instead of "if vAlive > 0," or "If vAlive > 0.9." The three conditions are identical for a boolean variable, and our computation should produce the same result in all three cases. All three conditions really mean $vAlive = 1$ or equivalently $vAlive \neq 0$. Our formula satisfies this property. Valuing $vAlive > x$ for any x in $(0, 1)$ leads to $e = vAlive - x$ with domain $\{-x, 1 - x\}$. Applying the formula, we get:

$$dt(e > 0) = \frac{a + e}{a + b} = \frac{vAlive - x - (-x)}{1 - x - (-x)} = vAlive$$

What about non-strict inequalities? Well, if a and b are both not 0, $e >= 0$ is identical to $e > 0$ and the same formula applies. If one of them is 0 and the inequality is not strict, then the condition is either always true or always false and this is a case we handle separately.

Finally, we consider the equality $e = 0$. Note $dt(e \neq 0) = 1 - dt(e = 0)$ so difference is covered, too. For a binary variable valued in $\{a, b\}$, we must have $a = 0$ or $b = 0$, otherwise it is always false. Say (without loss of generality) that $a = 0$ and $b > 0$. When e evaluates to 0, the equality is certainly true. Its DT is 1. When it evaluates to b, the DT of $e = 0$ is 0. In between, it is interpolated on a fuzzed e. Hence:

$$DT(e = 0) = 1 - \frac{e}{b}, DT(e \neq 0) = \frac{e}{b}$$

20.1.5 Discrete conditions

Finally, when $dom(e) = \{e_0, e_1, \ldots, e_n\}$ with $e_0 < e_1 < \ldots < e_n$, we say that the condition $e > [=]0$ is *discrete*. Discrete conditions include binary and therefore boolean conditions, so the formula for discrete conditions applies in all cases. We investigated only booleans and binaries separately for pedagogical reasons. The FCE does not apply, as we already demonstrated in particular cases. We may verify that the continuous formulas also fail with a more general discrete condition, like in the following:

Date1	If spot() < 100 then
	vLadder = 0
	else
	if Spot() < 110 then
	vLadder = 1
	else
	if spot() < 120 then
	vLadder = 2
	else
	if spot() < 130 then
	vLadder = 3
	else
	if spot() > 150 then
	vLadder = 5
	else
	vLadder = 4
	endIf
	endIf
	endIf
	endIf
	endIf

Date2	if vLadder > 3 then y = 1 else y = 0 endIf

In this script,[1] *vLadder* is a discrete variable with domain $\{0, 1, 2, 3, 4, 5\}$ and the fuzzy evaluator fuzzes it on date 1, so that when $vLadder > 3$ is evaluated on date 2, *vLadder* belongs to $(0, 5)$, and $e = vLadder - 3$ belongs to (-3.2). The domain of the expression $vLadder - 3$, however, is the discrete $\{-3, -2, -1, 0, 1, 2\}$ since the domains that matter are the ones deduced from a crisp reading.

With this in mind, we expect that $dt(e > 0)$ is 1 when $e >= 1$, 0 when $e <= 0$, and interpolated in between. More generally, when an expression e

[1]Part II, chapter 11 shows how this script could be written in a considerably less ugly form.

has a discrete domain $\{e_0, e_1, \ldots, e_n\}$, then fuzzy logic turns its domain into (e_0, e_n) and we compute the DT of $e > 0$ by interpolation:

$$\begin{aligned} &e \le e_{i_0} \rightarrow DT(e > 0) = 0 \\ &e \ge e_{i_1} \rightarrow DT(e > 0) = 1 \\ &e_{i_0} < e < e_{i_1} \rightarrow DT(e > 0) = \frac{e - e_{i_0}}{e_{i_1} - e_{i_1}} \end{aligned}$$

where e_{i_0} is the largest non-positive e_i, and $i_1 = i_0 + 1$. When we can't find such pair $\{e_{i_0}, e_{i_1}\}$ in $\{e_0, e_1, \ldots, e_n\}$, that means that the condition is always true or always false.

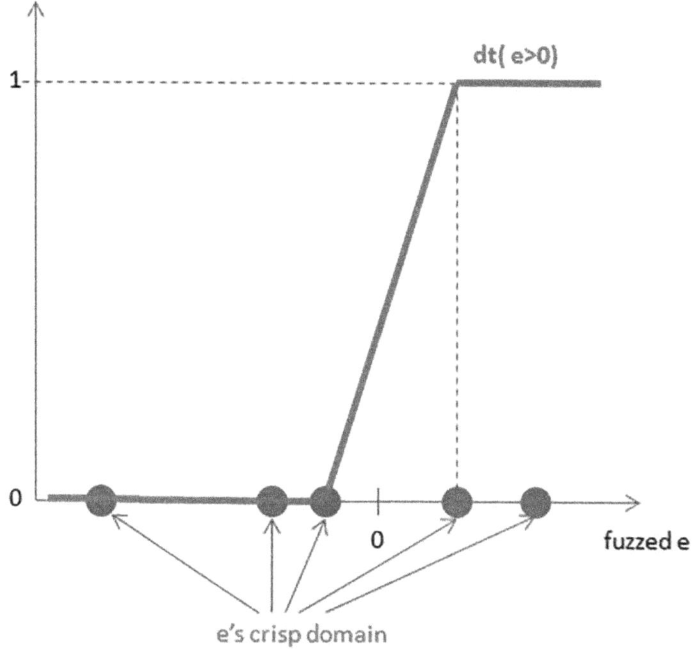

A relevant example is a counter of the number of periods the spot spends in a range:

W1	vCount = 0 if spot() > 90 and spot() < 110 then vCount = vCount + 1 endIf
W2	if spot() > 90 and spot() < 110 then vCount = vCount + 1 endIf
W3	if spot() > 90 and spot() < 110 then vCount = vCount + 1 endIf
W4	if spot() > 90 and spot() < 110 then vCount = vCount + 1 endIf if vCount > 1.5 then prd pays 100 endIf

On week 4 when *vCount* > 1.5 is evaluated, *vCount*'s crisp domain is $\{0, 1, 2, 3, 4\}$. We purposely wrote the condition as *vCount* > 1.5, which should be the same as *vCount* > 1 for example. We evaluate $e > 0$ with $e = vCount - 1.5$ and $dom(e) = \{-1.5, -0.5, 0.5, 1.5, 2.5\}$. Applying the formula for discrete DTs, we find that$\{e_{i_0}, e_{i_1}\} = \{-0.5, 0.5\}$ and

$$DT(e > 0) = 1_{\{-0.5<e<0.5\}}(e + 0.5) + 1_{\{e\geq 0.5\}}$$

So, if the spot spent a week or less in the range, then *vCount* <= 1, $e <= -0.5$, and $DT = 0$. If the spot spent 2 weeks or more in the range, $e > 0.5$, and $DT = 1$. Remember that *vCount* is fuzzed, hence continuous, when the script is evaluated with fuzzy logic. The fuzzed *vCount* belongs to $(0, 4)$, and our formula correctly interpolates DT when *vCount* is between 1 and 2. We can check that *vCount* > 1 produces the exact same result in all cases so the conditions evaluate consistently.

What if we have a non-strict condition, as in *vCount* >= 2, resolving into *vCount* – 2 >= 0? Well, the same formulas apply, except that in this case, we pick e_{i_0} as the largest strictly negative e_i. We can check that if we apply this formula with the previous script rewritten with *vCount* >= 2, we produce the same results in all scenarios.

Finally, equality conditions make sense with discretely valued conditions. In this case, we compute $DT(e = 0)$ with a butterfly $\{e_{i_{-1}}, 0, e_{i_1}\}$:

$$\begin{aligned}
e \leq e_{i_{-1}} &\rightarrow DT(e = 0) = 0 \\
e \geq e_{i_1} &\rightarrow DT(e = 0) = 1 \\
e_{i_{-1}} < e < 0 &\rightarrow DT(e = 0) = \frac{e - e_{i_{-1}}}{-e_{i_{-1}}} \\
0 < e < e_{i_1} &\rightarrow DT(e = 0) = \frac{e_{i_1} - e}{e_{i_1}}
\end{aligned}$$

where:

- $e_{i_{-1}}$ is the largest strictly negative singleton, or $-EPS$ when all singletons are positive.
- e_{i_1} is the smallest strictly positive singleton, or $+EPS$ when all singletons are negative.
- The singleton $\{0\}$ must be part of the value domain of e, otherwise the condition is always false. The case $dom(e) = \{0\}$ is also a particular case when the condition is always true.

Remarkably, there is no *smoothing factor* or spread size in smoothing discrete conditions; it is only for continuous conditions that we need an exogenous parameter.

20.1.6 Putting it all together

We finally know the formulas for the DT of all types of elementary conditions. The FCE algorithm correctly evaluates all condtional blocks provided that the correct formulas are used for DTs. These formulas also provide DTs of combined conditions (by the application of probabilistic style combinations) and nested conditions (the FCE automatically and implicitly multiplies the DTs of outer and inner conditions, consistently with how the *and* combinator evaluates).

In order to implement our formulas in the derived evaluator, we need to know, for all elementary conditions, their domains and whether they are always true or always false. This is to be determined by a visitor that pre-processes the script before simulations take place and stores domains and flicks always true/false flags on the elementary condition nodes. We design this visitor next. First, we summarize the correct formulas for the DT of all expressions knowing their domain and always true/false status.

1. If the condition is marked as *alwaystrue*, $dt = 1$.
2. If the condition is marked as *alwaysfalse*, $dt = 0$.
3. If the condition's domain is continuous, its $dt = fIf(e, 1, 0, EPS)$. Note that with continuous domains, equalities don't make sense and inequalities are all deemed strict.
4. If the condition domain is a discrete $\{e_0, e_1, \ldots, e_n\}$ such that the condition is not always true or false (otherwise it would have been caught earlier):
   (a) If the condition is a strict inequality of the form $e > 0$, find the pair $\{e_{i_0}, e_{i_1}\}$ in $\{e_0, e_1, \ldots, e_n\}$ such that e_{i_0} is the largest non-positive e_i, and $i_1 = i_0 + 1$, and apply:

$$
\begin{aligned}
e \le e_{i_0} &\rightarrow DT(e > 0) = 0 \\
e \ge e_{i_1} &\rightarrow DT(e > 0) = 1 \\
e_{i_0} < e < e_{i_1} &\rightarrow DT(e > 0) = \tfrac{e - e_{i_0}}{e_{i_1} - e_{i_1}}
\end{aligned}
$$

   (b) If the condition is a non-strict inequality of the form $e >= 0$, apply the same formula except that e_{i_0} is the largest strictly negative e_i.
   (c) If the condition is an equality of the form $e = 0$, then one of the e_i is 0 and there is at least another e_i that is not 0, otherwise the condition would have been caught as constant. Find $e_{i_{-1}}$ the largest

strictly negative e_i or −EPS if none, e_{i_1} the smallest strictly positive e_i or +EPS if none, and apply the butterfly formula:

$$\begin{aligned}
&e \leq e_{i_{-1}} \rightarrow DT(e = 0) = 0 \\
&e \geq e_{i_1} \rightarrow DT(e = 0) = 1 \\
&e_{i_{-1}} < e < 0 \rightarrow DT(e = 0) = \frac{e - e_{i_{-1}}}{-e_{i_{-1}}} \\
&0 < e < e_{i_1} \rightarrow DT(e = 0) = \frac{e_{i_1} - e}{e_{i_1}}
\end{aligned}$$

These formulas compute the DT of all types of elementary conditions given their domain and always true/false status. Everything else, including the computation of the DT of combined conditions and the propagation of fuzziness to affected variables, does not depend on condition domains. Only the DT of elementary conditions does. All the steps in the FCE remain correct as long as the DTs of all elementaries are computed with these formulas.

In order to complete the implementation, we need to compute the domains and check the true/false status of all elementary conditions. That is conducted with a dedicated visitor (actually two) at processing time.

20.2 IDENTIFICATION OF CONDITION DOMAINS

We now develop a pre-processor to compute all domains and store them on the elementary condition nodes (equality and inequality nodes) before simulations take place. Once the domains of all conditions are known, this visitor or another one can trivially identify the conditions that are always true or always false and flick the flags on their nodes accordingly. During simulations, the fuzzy evaluator has on the condition nodes the information it needs to correctly evaluate the DTs so the FCE can evaluate all conditional statements accordingly.

Note that there is no need to store the whole domain on the node. The only information the fuzzy evaluator needs is as follows:

1. Whether the condition is always true or always false
2. If not, whether the expression has continuous or discrete domain
3. If the domain is discrete, the two singletons surrounding 0 in the expression's domain, with strictness depending on the strictness of the condition

In order to compute this information, the domain processor must traverse in order all statements in the script and keep track of the domain of all variables as a state. The starting domain for all variables is the singleton $\{0\}$.

To avoid unnecessary complexity, we work under the assumption that some expression's domain is either continuous or discrete. In reality, an expression's domain may be the union of continuous intervals and singletons, as in the script:

```
        if spot() > 100
eDate   then z = spot() - 100
        else if spot() > 80 110 then z = -2 else z = -1 endIf
        endIf
```

The final domain for z is the union of $\{-2, -1\}$ and the positive real domain. This is, however, a twisted example. We may either call it practically irrelevant and ignore it as in what follows or handle it with a (substantially) more complicated visitor.

So domains are either represented as a discrete set of known numbers or are marked as continuous. The visitor keeps a vector of the domains of all variables as a state. It also maintains a stack of domains for the domains of sub-expressions/sub-trees. As usual, the computation takes advantage of the recursive nature of expression trees and the visitor framework. To compute the domain of an expression, we compute the domains of its arguments and push them onto its domain stack. Then we pull the argument's domains from the stack, apply the node's operation on domains to produce the node's domain, and push it back onto the stack. Operations on domains are conducted in exactly the same way that the evaluator conducts operations on numbers.

When the node being visited is an elementary condition, the domain of its LHS expression is used to store the necessary information on the condition's node—always true/false flags, and whether the domain is continuous and if not, the two singletons from the discrete domain that surround 0, and a strictness flag depending on the strictness of the condition. We remind again that at this stage, all elementary conditions are of the form *expression* > 0 or *expression* ≥ 0 or *expression* $= 0$. The domain of *expression* is recursively computed from the *expression*'s tree and the result is stored into the condition's top node, which is either a $>$, a $\geq$, or a $=$ node.

When we visit an assignment, the LHS variable's domain on the visitor's state is overwritten by the RHS expression's domain.

When we visit a payment, the LHS variable's domain is set to continuous, since the RHS expression is discounted by the numeraire, which may be valued in the positive real domain. Hence, anything that makes payment necessarily has a continuous domain.

When we visit leafs:

Constants obviously have domain $\{c\}$.

Variables domains are read from the visitor's state.

Simulated data like spot() have continuous domain.

When we visit functions, including operators:

- If one of the parameters has continuous domain, the function has continuous domain.
- Otherwise, the function has a discrete domain that is computed by the application of the function to all combinations of parameters where each parameter spans its discrete domain. If f is a function of three variables, for instance, its domain is the *list comprehension*, that is (in pseudo-code):

```
dom( f( arg1, arg2, arg3))
= { f( x,y,z) }
for x in dom( arg1), y in dom( arg2), z in dom( arg3)
```

Finally, when we hit a condition of the type

```
if [condition] then v=[expression] endIf
```

we want the domain of the affected variable v to be the union of whatever its domain was before the condition and the domain of *expression*. The domain of unaffected variables is obviously unchanged. If the domain of v was continuous, it remains continuous. If the domain of *expression* is continuous, the domain of v becomes continuous.

The domains of variables change throughout the script, which is why the domain processor must maintain the current domains of all variables as a state.

With a condition like "if Spot > 100 then x=1 else x=0 endIf," the resulting domain for x is obviously $\{0, 1\}$. But if we hit something like "if Spot > 100 then x=1 y = 1 else y = 0 z=0 endIf," things get more complicated: we add $\{1\}$ to the domains of x, we overwrite the domain of y with $\{0, 1\}$, and we add $\{0\}$ to the domain of z. The general algorithm for the processing of every conditional block comes out as follows:

The affected variables, that is variables that are written into, either by assignment or payment, in either the *iftrue* or *iffalse* statements, must have been previously identified by the affected variables' processor. Then:

1. Record the initial domains (D_0) of all affected variables.
2. Process *iftrue* statements and record resulting domains into (D_1).
3. Reset the domain of all affected variables to (D_0).
4. Process "if false" statements if any.
5. Record the domains resulting from the two previous steps into (D_2).
6. Set the final domains of all affected variables to the union of (D_1) and (D_2).

The visitor is not particularly fast; however, it runs only once, and in our tests, the time spent determining domains is always negligible compared to evaluation, including with large netting sets containing thousands of variables.

20.3 CONSTANT EXPRESSIONS

One side benefit with the domain processor is that it implicitly identifies constant expressions. Whenever the domain of an expression or sub-expression turns out to be a singleton $\{c\}$, this means that the expression is a constant, even when its expression tree is not a constant leaf, as in "100+10."

When a branch in an expression tree that is not already a constant leaf is identified as a constant expression, it is safe, and desirable to trim the branch by deleting its node in the expression tree and replacing it with a constant leaf that contains the constant value. "100+10" is replaced by "110," although that replacement is not conducted in the text, but in the tree, using a visitor.

We have not implemented the trimming of constants. If we had, we would have added a boolean *myIsConst* and a double *myConstVal* data members on the base *Node* object, initialized to false and 0. Whenever a constant domain (singleton $\{c\}$) is identified on the top of the domain stack during domain processing, the processor would flag the node being visited as constant by flicking its *myIsConst* and setting its *myConstVal* to *c*. Later, another processor (the "constant trimmer") would visit the trees, deleting all nodes flagged as constants and replacing them by constant leafs that contain their values.

What we have implemented is a "constant condition trimmer" that replaces all *if* nodes marked *alwaystrue* by the collection of its *istrue* statements and all *if* nodes market *alwaysfalse* by the collection of its *iffalse* statements.

CHAPTER 21
Limitations

We now have a fully working fuzzy evaluator that overrides the evaluation of conditional blocks resulting in an automatic smoothing that stabilizes the calculation of risk sensitivities with negligible impact on valuation and performance. As a necessary preamble, we implemented a domain processor that computes the domains of all elementary conditions and doubles up as a constant expression identifier/trimmer.

The implementation has been tested in many cases of practical relevance and provides, as expected, a substantial stabilization of risk sensitivities without a measurable performance impact or valuation bias. We can finally automatically smooth our simplified autocallable script, with the exact same results we achieved manually previously.

There are limitations, however, and we are investigating them now.

During extensive testing, we identified two families of limitations. Both relate to characteristics in the scripts that may be inconsistent with fuzzy logic, potentially causing a valuation bias or a crash. This is serious enough so that, at the very least, these limitations must be clearly identified and understood.

21.1 DEAD AND ALIVE

Consider the script:

eDate	if spot() > 100 then x = 1 else x= 0 endIf

Assume that the variable x was used previously in the script so on eDate it has some value, say 100.

eDate	x = 100 if spot() > 100 then x = 1 else x= 0 endIf

Fuzzy evaluation trivially and correctly results in

$$x = dt(Spot() > 100) = fIf(Spot() - 100, 1, 0, EPS)$$

But this statement could have been written as a combination of two statements:

```
          x = 100
eDate     if spot() > 100 then x = 1 endIf
          if spot() <= 100 then x = 0 endIf
```

In this case, we can verify that fuzzy evaluation results in

$$x = dt(Spot() > 100)^2 + 100dt(Spot() > 100)[1 - dt(Spot() > 100)]$$

(!) When the spot fixes above $100 + EPS$ or below $100 - EPS$ then $DT(spot() > 100) = 1$ or 0 and x correctly evaluates to 1 or 0. But when the spot fixes in between, we get the very wrong interpolation drawn here (EPS=1):

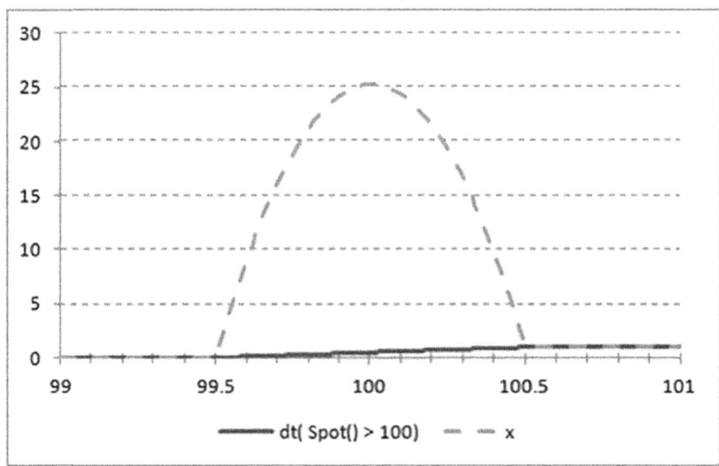

What happened here?

When the spot lands in the fuzzy region $(99.5, 100.5)$, conventional logic no longer applies. We find ourselves in a situation where, like the life and death of Schrodinger's cat, the conditions "spot() > 100" and "spot() < = 100" are both true, hence, the strange result.

The second script is identical to the first one only in appearance and only under the assumption that the conditions are exclusive. This assumption is trivially verified in conventional logic where A or not A = true but that does not hold with fuzzy logic. Unfortunately, sharp logic is hard-wired in our DNA, and such assumptions come naturally to us and may lead us to write scripts implicitly based on them that produce funny results with fuzzy logic.

Another example comes from our simplified autocallable. Recall the line:

```
01Jun2022    if spot() > vRef then
             if vAlive = 1 then prd pays 120 endIf
             vAlive = 0
             endIf
```

This works well with fuzzy logic. But we could have written instead:

```
01Jun2022    if vAlive = 1 and spot() > vRef
             then prd pays 120
             vAlive = 0
             endIf
```

This is more compact and at first sight identical. In both cases, we pay 120 provided that $vAlive = 1$ and $spot() > vRef$. The difference is that in the second script, the assignment $vAlive = 0$ is subject to *vAlive* being 1 to start with. This is not correct in principle. The rule is that the transaction dies if the spot fixes above the reference. But if the transaction was not alive to start with, then it was already dead, and there is no need to kill it again, right? Wrong. Not with fuzzy logic. If the transaction is *partly* dead from a former fixing inside the fuzzy region, it still needs further killing in part or in full when the spot fixes above the reference or in a fuzzy region around it. Again, the implicit reasoning behind the second script is that the deal has to be either alive or dead when we evaluate the event, so in case *vAlive* is already not 1, it has to be 0, and we can skip the assignment $vAlive = 0$. Such reasoning does not hold with fuzzy logic and produces wrong results.

We should stick to hard rules when writing scripts and absolutely refrain from introducing additional logic like "if spot did not fix above 100 then for sure it fixed below 100" or "if the transaction is not alive then for sure it is dead." Never subordinate scripts to the implicit assumption that (A and not A) is necessarily false, or (A or not A) is necessarily true. All that is no longer correct with fuzzy logic. Stick to the hard description of cash-flows and do not introduce logic that may no longer hold during fuzzy evaluation.

21.2 NON-LINEAR USE OF FUZZY VARIABLES

The second limitation we identified is more problematic. The rule is: fuzzed variables may not be safely used in non-linear expressions.

Fuzzed variables are those variables that are affected (assigned or paid into) in conditional blocks. For example, the following script fuzzes x into domain $(-1, 1)$:

```
Date1    if spot() > 100 then x = 1 else x = −1 endIf
```

After a fuzzy evaluation of the conditional block, it is okay to test these variables in conditions. We designed fuzzy evaluation accordingly. For example, say we receive a share if we fixed above 100, pay it if we fixed below:

```
Date2    if x = 1 then prd pays spot() else prd pays -spot() endIf
```

It is also OK to assign or pay an expression that involves fuzzed variables linearly. Remember that fuzzed variables are aggregates of their crisp values weighted by the part of the notional crisp values applied. To use them to multiply an expression means to apply the expression to the combined notional, which is perfectly valid. For example, we could have written:

```
Date2    prd pays x * spot()
```

Note that those variables that are assigned an expression involving fuzzed variables are themselves fuzzed in the process, which means that the same rules apply to them thereafter.

It is not okay to use fuzzed variables in a non-linear expression, either for test, assignment, or payment. For example:

```
Date2    prd pays spot() / x
```

This script works (miraculously) with sharp logic. With fuzzy logic, when the spot fixed close to 100, x is fuzzed to 0 and the system crashes.

While nobody in their sane mind would write this line, this limitation is quite restrictive in practice. Consider a European call where this strike depends on a previous fixing:

```
Date1    if spot() > 100 then k = 100 else k = 90 endIf
Date2    prd pays max( 0, spot() - k)
```

This script is legitimate, and yet it breaks the rule by using the fuzzed variable k in the non-linear expression $max(0, spot() - k)$. The script does not evaluate properly with fuzzy logic. In this case, we introduce a small bias, not a system-wide crash but to some extent this is even worse in finance. The script should be rewritten to comply with the rule:

```
Date1   if spot() > 100 then ki = 1 else ki = 2 endIf
          If ki = 1 then
Date2   pProduct pays max( 0, spot() - 100)
        else prd pays max( 0, spot() - 90)
        endIf
```

It is generally easy to rewrite those scripts that break the linearity rule manually. But it is unfortunately very hard to perform this transformation automatically. The least we can do is write a pre-processor that checks that the rule is enforced and, if broken, throws an exception inviting the user to rewrite the script. This is not implemented, and this whole non-linear story is subject for further research.

CHAPTER 22

The Smoothing Factor

In all that precedes, we consider as given the size EPS of the call spread we use for smoothing continuous conditions. In practice, the size of the call spread is generally set at a reasonable level by trial and error, or from commercial considerations (in which case the call spread is one-sided, not centered, so as to produce a positive bias used as a trading margin and a hedge buffer). This is not satisfactory. We know that the spread should be set low enough so as to minimize the valuation bias and large enough to produce sufficient smoothing. It should be determined automatically. As for using call spreads for commercial purpose, we disagree with this (very common) practice. We believe that the spread should be set for maximum efficacy and that the trading margin should be computed endogenously.

We did not succeed in producing an algorithm that would find the optimal spread in all cases, in the sense that it minimizes bias for a given simulation error or minimizes error for a given bias. Actually, we proved that it cannot be done, since such optimal spread depends on the context. For example, in the condition *spot* > 100, the optimal spread is different if that condition is used to determine the payoff of a digital or to discretely monitor a barrier. Since it is perfectly reasonable that this condition could do *both* in the same script, the notional of an optimal spread does not even make any sense.

That does not mean that we cannot do better than to use some arbitrary spread for all conditions. This chapter investigates what can be done.

22.1 SCRIPTING SUPPORT

Before we even move on to attempting the automatic determination of the spread, let us look at exactly how we can set it manually. We could pass a single spread as a parameter to the valuation function so that the fuzzy evaluator uses it to compute the DT of all continuous conditions. To value a script like:

```
01Jun2021     if spot() > 100 then dig pays 1 endIf
```

(today is 1 June 2020), we could pass 1 as a smoothing factor. To value another script like:

```
01Jun2021    if spot() / 100 > 1 then dig pays 1 endIf
```

we would need to pass another smoothing factor, certainly not 1; 0.01 is a much more reasonable choice. What if we have both conditions in the same script? In practice, it happens frequently that different conditions in the script require different spreads. We could have two shares trading about 20 and 5,000 or a share of order 100 and an interest rate of order 0.0025, and, very clearly, the same smoothing could not be applied to conditions involving one or the other. We could normalize all conditions so the expression in *expression > 0* is always of the same order of magnitude. But that means modifying scripts to suit algorithms, which is not satisfactory. We could design a visitor to perform this normalization automatically. But then, if we go that far, we may as well go all the way and design an algorithm to set the spread automatically for all conditions.

Either way, we need to provide a syntax for our scripting language to manually set the spread on individual elementary conditions. We suggest the following:

```
D1    vRef = spot()
      if spot() > 100 : 1 then prd pays 1 endIf
      vPerf = (spot() - vRef) / vRef
D2    if vPerf > 1 : 0.1 then prd pays 1 endIf
      vRef2 = spot()
      vPerf = (spot() - vRef2) / vRef2
D3    if vPerf > 1 : 0.1 and spot() > vRef : 2 then prd pays 1
      endIf
```

We possibly write a semicolon following each elementary condition with the spread to be used for smoothing. If this syntax is omitted on a particular condition, a default passed to the entry valuation function is used. This syntax allows us to manually specify spreads if we wish to do so or for testing our automatic algorithm. It is necessary to implement it in our scripting language but automatic determination is preferable. Note that even with automatic determination, we may want some specific conditions to use a user-defined spread instead. Our syntax may support such cases. We can write the code so that the spread is determined by the algorithm for conditions where the semicolon syntax is omitted and use the user-provided spread for conditions where it is explicitly present.

Our code in the online repository supports this syntax.

22.2 AUTOMATIC DETERMINATION

What is a sensible way to determine the smoothing spread for *expression* > 0?

One natural lead is to relate it to the standard deviation of *expression*. In practice, using EPS = .05 std(expression) generally produces satisfactory results with expressions about the money (which is expected to be close to 0).

It has been argued that out-of-the-money expressions should use a larger spread so that more paths fall inside the fuzzy region, improving smoothing. This means that we could choose the spread so that the proportion of paths landing in the fuzzy region $P(-EPS/2 < expression < EPS/2)$ is some target number, like 0.04 for consistency with .05 standard deviations at the money. We would need to evaluate the probability distribution for *expression* to implement this and that may be a costly or unstable procedure. Instead, we can only estimate the first two moments of *expression* and use a Gaussian approximation for the calculation.

This is all fine as long as we can estimate the moments of expressions involved in all elementary conditions. For this kind of computation, we would typically use pre-simulations. They consist of running a small number of Monte-Carlo simulations to pre-compute some quantities that are necessary for the main simulation before it takes place. It is typically used in modern implementations of the LSM algorithm to determine the coefficients to be used in the main simulation for the regression of PVs. Obviously, in case we need pre-simulations for a number of different purposes, we pre-simulate once and reuse the same pre-simulations to compute all the quantities we need.

For our purpose, we need pre-simulations to compute the moments of all elementary conditions and set the spread on their node accordingly. The generation of scenarios is not different from the main simulation. To evaluate them, however, we need a derived evaluator that, in addition to doing everything the evaluator does, keeps track of the realized values of the LHS expressions involved in all conditions for all scenarios. What we need to do is *decorate* the visitors to the elementary condition nodes so that (in addition to what they normally do) they write the realized values of their RHS expressions into the state of the derived evaluator. A pre-processor would also need to count and index conditions to start with and allocate space to hold the LHS expression values. When pre-simulations are complete, we would have all the realized values for the expressions involved in all conditions for all scenarios in the pre-simulation, so we can compute their moments, calculate the related spreads, and fire another visitor to write them on the condition nodes.

This logic has not been implemented in the code in the online repository. That code does not include pre-simulations.

This concludes our investigation of risk sensitivities for scripted transactions. We showed how to implement industry-standard smoothing in the language. We designed an algorithm that automatically smooths risks with a fuzzy logic evaluation of the script rather than a modification of the script. We designed a visitor that calculates the value domains of all conditions, necessary for the evaluation with fuzzy logic of discrete conditions. As a side benefit, this processor identifies conditions that are always true or always false and expressions that turn out to be constants. We investigated limitations to this approach and provided guidelines to deal with them. And, finally, we discussed the implementation of an automatic setting of smoothing spreads, introducing the important technique of pre-simulations. Putting it all together, we generalized market practice into new algorithms, which, combined, provide an automated, reliable, efficient smoothing for scripted transactions without the need to modify scripts or intervene manually in any way.

The identification of industry-wide smoothing for digitals, barriers, and other products as a particular use of fuzzy logic is in itself a very remarkable result. It is this realization that leads us to produce our algorithm, based on a derived evaluator that evaluates scripts in light of fuzzy logic in place of conventional logic.

The full source code can be found in our online repository.

PART V

Application to xVA

CHAPTER 23

xVA

We insisted many times that scripting supports xVA in a particularly practical and efficient manner. Looking into CVA without loss of generality, it is defined as follows (ignoring discounting):

$$CVA = E\left[\sum_i \left(1 - R_{T_i}\right) 1_{\{T_i \le \tau < T_{i+1}\}} \max\left(0, V_{T_i}\right)\right]$$

where τ is the default time, R the recovery rate, and V_{T_i} is the PV of the netting set at exposure time numer i, that is, the PV at that time of all future cash-flows from all the transactions in the netting set. We see that the language needs to cover credit so that we can script default and recovery. This may be implemented in a way that is identical to how we implemented asset prices: we may implement a forward survival probability keyword/node/request/data in a way identical to fwd for equities/commodities and recovery in a way identical to spots. In addition, xVA is often computed ignoring wrong-way risk, that is, with credit assumed independent from the drivers of the PV of the netting set. In this case, the formula simplifies into:

$$CVA = E\left[\sum_i \left[1 - R\left(0, T_i\right)\right]\left[S\left(0, T_i\right) - S\left(0, T_{i+1}\right)\right] \max\left(0, V_{T_i}\right)\right]$$

where R and S and deterministic recoveries and survival probabilities are read from today's credit markets. Since these quantities are deterministic in that case, they can simply be put on a predefined schedule—see part II, chapter 10—and xVA is supported without further developments.

Huge and Savine [18] report Andreasen's demonstration that CVA can also be written:

$$CVA = E\left[\sum_k \eta_{T_k} CF_k\right]$$

where CF_k is the cash-flow from the netting set paid on date T_k and

$$\eta_{T_k} = \sum_{T_i<T_k} \left(1 - R_{T_i}\right) 1_{\{T_{i-1}\leq\tau<T_i\}} 1_{\left\{V_{T_i}>0\right\}}$$

In other terms, to value a CVA means to value the underlying payments, discounted by a special path-dependent process:

$$\eta_0 = 0, \eta_{T_{i+1}} = \eta_{T_i} + \left(1 - R_{T_i}\right) 1_{\{T_{i-1}\leq\tau<T_i\}} 1_{\left\{V_{T_i}>0\right\}}$$

Or in the case of independent credit:

$$\eta_{T_{i+1}} = \eta_{T_i} + \left[1 - R\left(0, T_i\right)\right] \left[S\left(0, T_i\right) - S\left(0, T_{i+1}\right)\right] 1_{\left\{V_{T_i}>0\right\}}$$

Hence, provided that all the transactions in the netting set are correctly scripted, aggregated, and compressed,[1] we have all cash-flows in the netting paid on a variable *ns* in a single (generally long) script:

CFDate1	ns pays ... ns pays ... ... ns pays ...
CFDate2	ns pays ... ns pays ... ... ns pays ...
...	...
CFDateN	ns pays ... ns pays ... ... ns pays ...

[1]Which is also done by a dedicated visitor; see Andreasen's CVA on an iPad Mini Part 3 [1], slides 26 and up.

All we need in order to compute the CVA (or any other xVA) is to *decorate* this script, that is, keep it as is but add the following:

1. A duplicate of all payments on a variable called cva, but multiplied by nu. This is conducted by a dedicated visitor:

...	...
CFDateI	ns pays ... ns pays ... ... ns pays ... cva pays nu * ... cva pays nu * ... ... ns pays nu * ...
...	...

2. The script for

$$\eta_{T_{i+1}} = \eta_{T_i} + \left[1 - \mathrm{R}(0, T_i)\right] \left[S\left(0, T_i\right) - S\left(0, T_{i+1}\right)\right] 1_{\left\{V_{T_i} > 0\right\}}$$

Start: TODAY	if PV(ns) > 0 then
End: FINALDATE	nu = nu + (1 - rec(StartPeriod))
Freq: CVAFREQ	* (surv(StartPeriod) - surv(EndPeriod))
Fixing: end	endIf

And just like that, we have a script for the CVA (or any other xVA in a similar manner). Of course, we need a hybrid model to properly value this script.

CHAPTER 24

Branching

Perhaps surprisingly, Andreasen's CVA reformulation does not extend by itself to the collateralized case, where the CVA is defined as:

$$CVA = E\sum \left(1 - R_{T_i}\right) 1_{\{T_i \le \tau < T_{i+1}\}} \max\left(0, V_{T_i} - V_{T_i - T}\right)$$

where T is the margin period of risk (MPR). Huge and Savine [18] show that this case cannot be resolved in the context of standard Monte-Carlo simulations. It is necessary to have recourse to so-called *branching simulations*. Branching Monte-Carlo simulations were recently produced by Henry-Labordere [17], under the inspiration of 40-year-old work by McKean [23]. Andreasen and Huge successfully applied this technique to resolve the problem of the collateralized CVA. Subsequently, systematic support for branching was implemented in Danske Bank's CVA systems by Savine and extended to produce RWA and KVA without nested simulation by Flyger, as was demonstrated live in Global Derivatives in 2015 and 2016 by computing full-risk sensitivities for an RWA on a large netting set locally on MacBook Pro within a couple of minutes [11].

Branching Monte-Carlos consist of the simulation, in every scenario, of not only one evolution of the world but also a number of "branches," where each branch sticks out of the main simulation at a particular point of time (for our purpose, the collateral fixing dates) and starts a parallel simulation that follows the same dynamics as the "main branch" but independently from it:

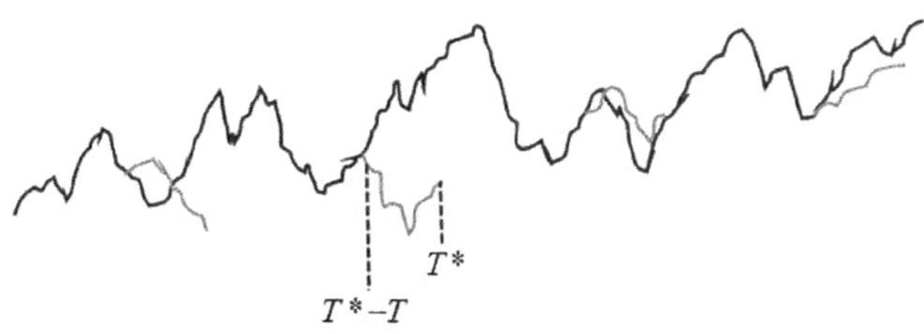

Given that our Monte-Carlo simulates some multidimensional diffusion for the state variables of the model

$$d\vec{S} = a\left(\vec{S}, t\right) dt + b\left(\vec{S}, t\right) dW$$

a secondary branch sticking out at time T_B, noted ${}^B\vec{S}$, is part of the same scenario and simulated in such a way that:

1. For $t \le T_B \; {}^B\vec{S}^s_t = \vec{S}^s_t$.
2. $t > T_B d^B\vec{S} = a\left({}^B\vec{S}, t\right) dt + b\left({}^B\vec{S}, t\right) dW^B$ where W^B is a standard multidimensional Brownian Motion *independent from* W.

Those "hairy simulations" are different from "nested" simulations. We simulate only one hair for every branch, and the hairs are part of the scenario along with the central branch. Then, if we denote ${}^BV_T$ the PV at time T simulated on the branch B, Huge and Savine show that the formula for the uncollateralized CVA holds:

$$CVA = E\left[\sum_k \eta_{T_k} CF_k\right]$$

except now:

$$\eta_{T_k} = \sum_{T_i < T_k} \left(1 - R_{T_i}\right) 1_{\{T_{i-1} \le \tau < T_i\}} \left(1_{\left\{V_{T_i} > V_{T_i - T}\right\}} - 1_{\left\{{}^BV_{T_i} > V_{T_i - T}\right\}}\right)$$

In other terms:

$$\eta_0 = 0, \eta_{T_{i+1}} = \eta_{T_i} + \left(1 - R_{T_i}\right) 1_{\{T_{i-1} \le \tau < T_i\}} \left(1_{\left\{V_{T_i} > V_{T_i - T}\right\}} - 1_{\left\{{}^BV_{T_i} > V_{T_i - T}\right\}}\right)$$

Or in the case of independent credit:

$$\eta_{T_{i+1}} = \eta_{T_i} + \left[1 - R\left(0, T_i\right)\right] \left[S\left(0, T_i\right) - S\left(0, T_{i+1}\right)\right] \left(1_{\left\{V_{T_i} > V_{T_i - T}\right\}} - 1_{\left\{{}^BV_{T_i} > V_{T_i - T}\right\}}\right)$$

The support for branching in the simulation library may be implemented in a model-agnostic way, provided all models satisfy an API for producing partial paths given an initial set of state variables. The scripting library must also be extended to support branching. This rather advanced development, which was carried in full by Savine in Danske Bank, is not covered in this book. Its result is the ability to script collateralized xVA is follows:

```
Start: TODAY
End: FINALDATE      collat = PV( ns)
Freq: CVAFREQ       branchOut( 1)
Fixing: end-MPR
                    V0 = PV( ns)
Start: TODAY        onBranch 1 VB = PV( ns) endBranch
End: FINALDATE      if V0 > collat and VB < collat then
Freq: CVAFREQ       = nu + ( 1 - rec( StartPeriod) )
Fixing: end         = * ( surv( StartPeriod) - surv( EndPeriod) )
                    endIf
```

The keyword *branchOut* is a marker that tells the simulator to start a hair at that event date and give it some name or index. The statements within *onBranchidx* and *endBranch* evaluate on the hair *idx*. That means that the evaluator works as usual, except the simulated values: spot, fwd, and so on, including state variables, are picked on that branch. A dedicated branch visitor identifies after what date a branch is no longer used, so there is no need to manually end a branch with a keyword like *terminateBranchidx*.

CHAPTER 25

Closing Remarks

We described all the steps involved in the production and use of a professional scripting language in modern C++. We provided a structure based on visitable expression trees, so that visitors access the scripted cash-flows for evaluation against simulations and much more. We insisted on the development of specialized visitors, like pre-processors and indexers, which improve the efficiency of evaluation so its run time is similar to optimized, hard-coded evaluation code. As an example for what may be done with scripting aside from valuation, we showed how this structure lends itself to evaluation with fuzzy logic to automatically stabilize risk sensitivities. Visitors can "peek" into the "source code" for the cash-flows, so they can do an amazing number of things, ranging from optimizations to automatic smoothing to aggreageation and compression of transactions and much more.

It is not our purpose to deliver a finalized software, but to demonstrate how it can be implemented. For that reason, we exposed C++ source code in the body of the document for the fundamental components of the scripting library. The entire source is available in our repository:

https://github.com/asavine/Scripting/tree/Book-V1

(branch Book V_1, not master). We also provide the source code fuzzy logic there. We do not provide source for the professional improvements and extensions discussed in parts II and III. This is for the readers to implement in accordance to their needs and the constraints of their infrastructure and environment. The explanations given in the book should hopefully enable them to perform these developments seamlessly over the backbone developed in part I.

25.1 SCRIPT EXAMPLES

We collect here a few examples of scripts that demonstrate the finalized version of the scripting language.

Equity Call

STRIKE	120
MATURITY	01Jun2021
CALL(S,K)	max(0, S-K)
MATURITY	opt pays CALL (Spot(), STRIKE)

Forex Barrier

INDEX	fx(EUR, USD)
STRIKE	100
BARRIER	120
PAYOFF(S, K)	max(0, S-K)
MATURITY	01Jun2021
BARSTART	01Jun2020
BAREND	01Jun2021
BARFREQ	1m
BARSTART	vAlive = 1
Start: BARSTART End: BAREND Freq: BARFREQ Fixing: end	if INDEX > BARRIER then vAlive = 0 endIf
MATURITY	opt pays vAlive * PAYOFF(INDEX, STRIKE)

Basket Option

TRADEDATE	01Jun2020
EXPIRY	01Jun2021
STRIKE 0.1	
STOCKS	WEIGHTS
STOCK1	0.2
STOCK2	0.1
STOCK3	0.15
STOCK4	0.25
STOCK5	0.3

TRADEDATE	loop (0, NSTOCKS) s0[ii] = spot(stocks[ii]) endLoop
EXPIRY	loop (0, NSTOCKS) s1 = spot(stocks[ii]) perf = perf + weights[ii] * (s1 - s0[ii]) / s0[ii] endLoop opt pays max(0, perf - STRIKE)

Interest Rate Swap

STARTDATE	01Jun2020
ENDDATE	01Jun2030
FLFREQ	3m
FLBASIS	act/360
FLIDX	L3M
FIXFREQ	1y
FIXBASIS	30/360
CPN	2%
Start: STARTDATE End: ENDDATE Freq: FLFREQ Fixing: start-2bd	swap pays -libor(StartPeriod, EndPeriod, FLBASIS, FLIDX) * cvg(StartPeriod, EndPeriod, FLBASIS) on EndPeriod
Start: STARTDATE End: ENDDATE Freq: FIXFREQ Fixing: start-2bd	swap pays CPN * cvg(StartPeriod, EndPeriod, FIXBASIS) on EndPeriod

Cap

STARTDATE	01Jun2020
ENDDATE	01Jun2030
FLFREQ	3m
FLBASIS	act/360
FLIDX	L3M
STRIKE	1.5%
Start: STARTDATE End: ENDDATE Freq: FLFREQ Fixing: start-2bd	cap pays max(0, libor(StartPeriod, EndPeriod, FLBASIS, FLIDX) -STRIKE) * cvg(StartPeriod, EndPeriod, FLBASIS) on EndPeriod

Interest Rate Swap with Capped Coupons

STARTDATE	01Jun2020
ENDDATE	01Jun2030
FLFREQ	3m
FLBASIS	act/360
FLIDX	L3M
FIXFREQ	1y
FIXBASIS	30/360
CPN	1.5%
CPN	2%
Start: STARTDATE End: ENDDATE Freq: FLFREQ Fixing: start-2bd	swap pays -min(CAP, libor(StartPeriod, EndPeriod, FLBASIS, FLIDX)) * cvg(StartPeriod, EndPeriod, FLBASIS) on EndPeriod
Start: STARTDATE End: ENDDATE Freq: FIXFREQ Fixing: start-2bd	swap pays CPN * cvg(StartPeriod, EndPeriod, FIXBASIS) on EndPeriod

Swaption

TRADEDATE	01Jun2020
STARTDATE	01Jun2020
ENDDATE	01Jun2030
FLFREQ	3m
FLBASIS	act/360
FLIDX	L3M
FIXFREQ	1y
FIXBASIS	30/360
CPN	2%
Start: STARTDATE End: ENDDATE Freq: FLFREQ Fixing: start-2bd	if vAlive = 1 then swaption pays -libor(StartPeriod, EndPeriod, FLBASIS, FLIDX) * cvg(StartPeriod, EndPeriod, FLBASIS) on EndPeriod endIf

<table>
<tr><td>Start: STARTDATE
End: ENDDATE
Freq: FIXFREQ
Fixing: start-2bd</td><td>if vAlive = 1 then swaption pays
CPN * cvg(StartPeriod, EndPeriod, FIXBASIS)
on EndPeriod</td></tr>
<tr><td>TRADEDATE</td><td>vAlive = 1</td></tr>
<tr><td>STARTDATE-2BD</td><td>terminate vAlive when PV(swaption) < 0</td></tr>
</table>

Bermuda

<table>
<tr><td>TRADEDATE
STARTDATE
ENDDATE
FLFREQ
FLBASIS
FLIDX
FIXFREQ
FIXBASIS
CPN</td><td>01Jun2020
01Jun2020
01Jun2030
3m
act/360
L3M
1y
30/360
2%</td></tr>
<tr><td>Start: STARTDATE
End: ENDDATE
Freq: FLFREQ
Fixing: start-2bd</td><td>if vAlive = 1 then swaption pays
-libor(StartPeriod, EndPeriod, FLBASIS,
FLIDX)
* cvg(StartPeriod, EndPeriod, FLBASIS)
on EndPeriod
endIf</td></tr>
<tr><td>Start: STARTDATE
End: ENDDATE
Freq: FIXFREQ
Fixing: start-2bd</td><td>if vAlive = 1 then swaption pays
CPN * cvg(StartPeriod, EndPeriod, FIXBASIS)
on EndPeriod</td></tr>
<tr><td>TRADEDATE</td><td>vAlive = 1</td></tr>
<tr><td>Start: STARTDATE+2Y
End: ENDDATE
Freq: FIXFREQ
Fixing: start-10bd</td><td>terminate vAlive when PV(swaption) < 0</td></tr>
</table>

Exotic Path-Dependent Bermuda

TRADEDATE STARTDATE ENDDATE FLFREQ FLBASIS FLIDX FIXFREQ FIXBASIS CPN	01Jun2020 01Jun2020 01Jun2030 3m act/360 L3M 1y 30/360 2%
Start: STARTDATE End: ENDDATE Freq: FLFREQ Fixing: start-2bd	c = libor(StartPeriod, EndPeriod, FLBASIS, FLIDX) maxCpn = max(c, maxCpn) if vAlive = 1 then swaption pays -maxCpn * cvg(StartPeriod, EndPeriod, FLBASIS) on EndPeriod endIf
Start: STARTDATE End: ENDDATE Freq: FIXFREQ Fixing: start-2bd	if vAlive = 1 then swaption pays CPN * cvg(StartPeriod, EndPeriod, FIXBASIS) on EndPeriod
TRADEDATE	vAlive = 1
Start: STARTDATE+2Y End: ENDDATE Freq: FIXFREQ Fixing: start-10bd	terminate vAlive when PV(swaption) < 0

25.2 MULTI-THREADING AND AAD

Our code correctly templates active numbers so it is ready for AAD instrumentation. In fact, the code has been successfully tested with AAD.

It is also thread-safe so simulations can be multi-threaded across simulations. Of course, each thread must use its own copy of the evaluator, since the evaluator's state changes during simulations. But the evaluator (and its derived versions) is the only object that writes into its state during simulations. The rest of the scripting library is safe.

Of course, the model must be thread-safe itself. For instance, the simple implementation of Black-Scholes we used in part I, chapter 4 to test run our library is not thread-safe. We can still multi-thread simulations but each thread needs its own copy of the model.

25.3 ADVANCED LSM OPTIMIZATIONS

Finally, it was mentioned in part III, chapter 15 that the support for LSM could be massively optimized by the selective evaluation of the events that affect the regressed variable, directly or indirectly. In order to implement that important optimization, a complete dependency graph of variables to other variables must be established. That may be conducted with a dedicated visitor that provides the list of all variables on which the regressed variable depends directly or not, including itself. Once we have that list, it easy to evaluate only statements that write into variables on that list and ignore the others. The performance impact may be very significant, literally from hours to less than a minute for large CVAs.

In addition, that dependency information may materially speed up AAD risk. AAD complexity is constant in the number of inputs, but linear in the number of outputs. Therefore we typically don't compute risks for all variables in a script, but only for one or a limited number of variables that we specify. There is no need to evaluate events that don't affect those variables, directly or not, especially in the context of AAD where computations typically take four to eight times longer (see one of the numerous publications or lectures on AAD, such as Naumann's [24]).

The dependency processor makes a significant difference in the performance of LSM and AAD. Its implementation is not covered in this book (just because we can't cover everything). It has been implemented by Savine in Danske Bank's award-winning systems (the In House System of the Year 2015 Risk award was awarded to Danske Bank for its CVA system, and received by Andreasen and Savine on behalf of the Bank).

APPENDIX A
Parsing

A.1 PREPARING FOR PARSING

Before we parse a string, we need to *tokenize* it, that is, turn something like

if max(spot(), vAverage) < 120 then prd = 1 else prd = 0

into

```
{
   "IF", "MAX", "(", "SPOT", "(", ")", "VAVERAGE",
   ")", "<", "120", "THEN", "PRD", "=", "1",
   "ELSE", "PRD", "=", "0"
}
```

We conveniently reuse the C++11 regular expression (or regex) standard library. This is a powerful and versatile string manipulation library that is part of the C++11 standard. This library is very useful but it comes with a somewhat steep learning curve. To use it for tokenizing is a bit of overkill, but then we get a fully functional tokenizer with just a few lines of code. We refer to a C++11 textbook for an overview. Our tokenizer splits a string into words, numbers, tokens

```
{(, ), +, -, *, /, ^, =, !=,<, >,<=,>=}
```

and commas. It also converts all letters to uppercase. It skips whitespaces and other tabs, newlines, and so on.

```
#include <regex>
#include <algorithm>

vector<string> tokenize( const string& str)
{
  // Regex matching tokens of interest
  static const regex r
     ( "[\\w.]+|[/-]|,|[\\(\\)\\+\\*\\^]|!=|>=|<=|[<>=]");
```

```
    // Result, with max possible size reserved
    vector<string> v;
    v.reserve( str.size());

    // Loop over matches
    for( sregex_iterator it
        ( str.begin(), str.end(), r), end; it != end; ++it)
    {
      // Copy match into results
      v.push_back( (*it)[0]);
      // Uppercase
      transform
          ( v.back().begin(), v.back().end(), v.back().begin(), toupper);
    }

    // C++11 move semantics means no copy
    return v;
}
```

Next step is to parse the sequence of tokens into trees. On the highest level of the parsing process, we add a method in our *Product* class to build the whole product from a number of event strings. We pass the event dates and strings as iterators on pairs of dates and strings, for instance, as given by the *begin*() and *end*() methods of a map of dates to strings. These are better data structures for scripts than, say, a vector of dates and a corresponding vector of strings. However, internally to the *Product* where information is encapsulated, we would rather hold two separate vectors, because the model needs access to the event dates for simulations, but never accesses the events directly.

```
class Product
{
    vector<Date>   myEventDates;
    vector<Event>  myEvents;

// ...
public:

    // ...

    // Build events out of event strings
    template<class EvtIt>
    // Takes begin and end iterators on pairs of dates
    // and corresponding event strings
    //  as from a map<Date,string>
    void parseEvents( EvtIt begin, EvtIt end)
    {
      // Copy event dates and parses event strings sequentially
      for( EvtIt evtIt = begin; evtIt != end; ++evtIt)
```

```
    {
      // Copy event date
      myEventDates.push_back( evtIt->first);
      // Parse event string
      myEvents.push_back( parse( evtIt->second));
    }
  }
};
```

And we parse each event string with a call to our parent parsing function:

```
// Event = vector<Statement> = vector<ExprTree>
Event parse( const string& eventString)
{
  Event e;

  auto tokens = tokenize( eventString);

  auto it = tokens.begin();
  while( it != tokens.end())
  {
    e.push_back( Parser<decltype(it)>::
       parseStatement( it, tokens.end()));
  }

  // C++11 --> vectors are moved, not copied
  return e;
}
```

This event parser first tokenizes the event string, then initializes a token iterator *it* to the first token and calls the statement parser in a loop. The statement parser parses exactly one statement that begins with the *it* token. It returns the (top node of) the resulting expression (sub-)tree and advances it to the token that immediately follows the last token in the statement. We record the result in the event—we remind that events are vectors of trees, one per statement—and loop until we reach the end of the tokens for the event. So a syntax like "x = 0 y = 0" produces two trees in an event.

Note the (somewhat obscure) syntax:

```
e.push_back( Parser<decltype(it)>::
parseStatement( it, tokens.end()));
```

We chose to wrap all parsing functions as static methods in a *Parser* class templated for token iterators (in case we may want to use something else than a vector of strings to hold tokens). This class wraps all the parsing functions that we develop in this section as private static methods and only publicly exposes the top level *parseStatement*(). Hence, to parse each statement, we call *parseStatement*() on the *Parser* class templated by the type of our iterator *it*.

A.2 PARSING STATEMENTS

A statement is characterized by an instruction. We have two instructions so far in our simplified syntax: the form

```
if condition then statements [else statements] endIf
```

(that itself has nested statements) and the assignment

```
var = expression
```

We will add more as we extend the syntax. For now, we note that every statement is built around an *instruction*. An expression without an instruction, say "x*y," is not a fully formed statement and we cannot work with it. Hence, the first task of the statement parser is to *identify* the instruction. Instructions come in two flavors: the ones that start with their identifier, like "if...," and the ones that start with their left-hand side (LHS) argument, like the "LHS = RHS" assignment with their identifier (in this case "=") somewhere in the core of the statement.

Operators that take LHS and RHS arguments, and their precedence rules, are the only reason parsers are somewhat hard. If the language was only made of functions, for instance, if we would write

```
if(sup(a,b),assign(c,add(d,e)))
```

instead of

```
if a>b then c=d+e endIf
```

then we would be done in half a page. But the end result would not be very attractive, and scripting languages would probably not have had such success with derivatives professionals.

A simplifying rule that may not hold in more sophisticated languages like C++, but is good enough for financial scripting, is that the LHS in an instruction of the second type is always a variable.

The first instruction type is easily identified by its first token. If we have a match, we call a dedicated parsing function for the corresponding instruction. If not, we parse the first token as a variable and identify the instruction with the second token. If we still don't have a match, then our statement is not well formed and we have an error.

```
struct script_error : public runtime_error
{
  script_error( const char msg[]) : runtime_error( msg) {}
};

// Statement = ExprTree = unique_ptr<Node>
static Statement parseStatement( TokIt& cur, const TokIt end)
{
  // Check for instructions of type 1, so far only 'if'
  if( *cur == "IF") return parseIf( cur, end);

  // Parse cur as a variable
  auto lhs = parseVar( cur);

  // Check for end
  if( cur == end) throw script_error( "Unexpected end of statement");

  // Check for instructions of type 2, so far only assignment
  if( *cur == "=") return parseAssign( cur, end, lhs);

  // No instruction, error
  throw script_error( "Statement without an instruction");
  return Statement();
}
```

Now we must implement three parsers to cover all cases: one for *if* instructions, one for variables, and one for assignments. We start with *parseIf*(). In an *if* instruction, the "if" token is followed by a condition, itself followed by the "then" token, followed by a number of statements, followed by either "endIf" or "else"; and in case it is "else," we have another series of statements and finally "endIf." The statements following "then" and "else" are parsed with recursive calls to *parseStatement*(), because they are full statements themselves; in particular they may contain nested *if* instructions. We remind that the *if* node holds the index of the first statement that follows "else" or −1 if we have no "else." Implementation follows.

```
static ExprTree parseIf( TokIt& cur, const TokIt end)
{
  // Advance to token immediately following "if"
  ++cur;

  // Check for end
  if( cur == end)
    throw script_error( "'If' is not followed by 'then'");

  // Parse the condition
  auto cond = parseCond( cur, end);

  // Check that the next token is "then"
  if( cur == end || *cur != "THEN")
    throw script_error( "'If' is not followed by 'then'");
```

```
// Advance over "then"
++cur;

// Parse statements until we hit "else" or "endIf"
vector<Statement> stats;
while( cur != end && *cur != "ELSE" && *cur != "ENDIF")
  stats.push_back( parseStatement( cur, end));

// Check
if( cur == end)
  throw script_error
      ( "'If/then' is not followed by 'else' or 'endIf'");

// Else: parse the else statements
vector<Statement> elseStats;
int elseIdx = -1;
if( *cur == "ELSE")
{
  // Advance over "else"
  ++cur;
  // Parse statements until we hit "endIf"
  while( cur != end && *cur != "ENDIF")
    elseStats.push_back( parseStatement( cur, end));
  if( cur == end)
    throw script_error
       ( "'If/then/else' is not followed by 'endIf'");
  // Record else index
  elseIdx = stats.size() + 1;
}

// Finally build the top node
auto top = make_node<NodeIf>();
top->arguments.resize( 1 + stats.size() + elseStats.size());
top->arguments[0] = move( cond);
   // Arg[0] = condition
for( size_t i=0; i<stats.size(); ++i)
   // Copy statements, Arg[1..n-1]
  top->arguments[i+1] = move( stats[i]);
for( size_t i=0; i<elseStats.size(); ++i)
// Copy else statements, Arg[n..N]
  top->arguments[i+elseIdx] = move( elseStats[i]);
top->firstElse = elseIdx;

// Advance over endIf and return
++cur;
return move( top);
// Explicit move is necessary
// because we return a base class pointer
}
```

Note how the parser builds the tree recursively. First, its builds the tree for the condition with a call to *parseCond*() and holds the address of its top node in the pointer *cond*. Next, we build trees for all statements that come after "then" and "else" and hold their top nodes addresses in pointers.

This is done with recursive calls to *parseStatement*(). Finally, we build the top *if* node, make *cond* its first argument indexed 0 and the top node of parsed statements its arguments 1 through *n*. We record the index of the first statement after "else" in the node if necessary, −1 otherwise, and return a pointer to the top node. When the method returns the top *if* node, the full tree beneath has been constructed, including possible nested *if* trees thanks to the recursive calls to *parseStatement*(), which itself calls *parseIf*() should it run into another nested "if" token.

Now we need to implement a parser for conditions, *parseCond*(), but first we implement the simpler *parseVar*() and *parseAssign*().

```
static ExprTree parseVar( TokIt& cur)
{
  // Check that the variable name starts with a letter
  if( (*cur)[0] < 'A' || (*cur)[0] > 'Z')
    throw script_error
    ( (string( "Variable name ") + *cur + " is invalid").c_str());

  // Build the var node
  auto top = make_node<NodeVar>();
  top->name = *cur;

  // Advance over var and return
  ++cur;
  return move( top);
  // Explicit move is necessary
  // because we return a base class pointer
}
```

The code above is self-explanatory. The reason for the explicit *rValue* cast in the return statement is that we built a concrete node pointer and return it as a base node pointer. In this case, the compiler will not automatically move the result, and it must be done manually, since unique_ptr is a move-only type. Moving to *parseAssign*():

```
static ExprTree parseAssign( TokIt& cur, const TokIt end, ExprTree& lhs)
{
  // Advance to token immediately following "="
  ++cur;

  // Check for end
  if( cur == end) throw script_error
     ( "Unexpected end of statement");

  // Parse the RHS
  auto rhs = parseExpr( cur, end);

  // Build and return the top node
  return buildBinary<NodeAssign>( lhs, rhs);
}
```

Here, *buildBinary*() is a helper that builds a node of the type corresponding to the template argument, sets LHS and RHS as its arguments, and returns a base Node (smart) pointer to the newly built node:

```
template <class NodeType>
static ExprTree buildBinary( ExprTree& lhs, ExprTree& rhs)
{
  auto top = make_base_node<NodeType>();
  top->arguments.resize( 2);
  // Take ownership of lhs and rhs
  top->arguments[0] = move( lhs);
  top->arguments[1] = move( rhs);
  // Return
  return top;
}
```

The assignment parser *parseAssign*() is passed a pre-constructed LHS so it only needs to parse the RHS. In an assignment, the RHS is an expression, that is, something that is neither an instruction nor a condition. Expressions are sequences of operators, functions, constants, and variables that evaluate to a number. We call a dedicated expression parser *parseExpr*() that parses one such expression in full. Then, in *buildBinary*(), we build the top assign node, make the LHS variable its first argument and the parsed RHS expression its second argument, and return it. When execution completes, the whole tree for the instruction "var = expression" is built and its top node is returned. The iterator *cur* advanced to the first token *following* the statement, or ends if there are no more tokens in the event.

This leaves us with two parsers to code: the condition parser *parseCond*() and the expression parser *parseExpr*(). We will see that the two are very similar in their structure. We start with the condition parser, which has fewer operators and is therefore marginally easier, in an attempt to expose the key concepts in a somewhat simpler context.

A.3 RECURSIVELY PARSING CONDITIONS

A condition may be elementary as in "expression comparator expression," for example, "x > y." These are parsed as follows: first we parse the LHS expression, next we identify the comparator among =, ! =, <, >, <= and >=, then we parse the RHS expression, and finally we build the top node according to the comparator, make LHS and RHS its arguments, and return it. Implementation follows.

```
static ExprTree parseCondElem( TokIt& cur, const TokIt end)
{
  // Parse the LHS expression
  auto lhs = parseExpr( cur, end);
```

```
    // Check for end
    if( cur == end) throw script_error( "Unexpected end of statement");

    // Advance to token immediately following the comparator
    string comparator = *cur;
    ++cur;

    // Check for end
    if( cur == end) throw script_error( "Unexpected end of statement");

    // Parse the RHS
    auto rhs = parseExpr( cur, end);

    // Build the top node, set its arguments and return
    if( comparator == "=")
        return buildBinary<NodeEqual>( lhs,rhs);
    else if( comparator == "!=")
        return buildBinary<NodeDifferent>( lhs,rhs);
    else if( comparator == "<")
        return buildBinary<NodeInferior>( lhs,rhs);
    else if( comparator == ">")
        return buildBinary<NodeSuperior>( lhs,rhs);
    else if( comparator == "<=")
        return buildBinary<NodeInfEqual>( lhs,rhs);
    else if( comparator == ">=")
        return buildBinary<NodeSupEqual>( lhs,rhs);
    else
        throw script_error( "Elementary condition has no valid comparator");
}
```

But not all conditions are elementary. Conditions may combine an arbitrary number of elementary conditions with *and* and *or*. In addition, we must respect the conventional precedence of *and* over *or*, so that $c1andc2orc3andc4$ reads $(c1andc2)or(c3andc4)$, while allowing users to supersede precedence with parentheses as in $c1and(c2orc3)andc4$. This is where we need the multi-level recursive descent.

Think of a compound condition like "X is a feline or X is a canine and X is not a dog," and how you would proceed to resolve it (decide if it is true or false) for a given X (say a wolf). First you would resolve all the elementary conditions involved (feline = false, canine = true, not a dog = true). This leaves you with "false or true and true." Elementaries must be resolved first, which we rephrase as "elementaries have highest precedence." Next you would resolve the and combinations, since they conventionally precede or combinations, so your "false or true and true" should read "false or (true and true)"; true and true resolves into true; this leaves you with "false or true." Finally, you evaluate the lowest precedence *or* combinator: false or true = true; so finally the whole condition resolves to *true*.

Scripted conditions are processed in the same way. In "c1 and c2 or c3 and c4," we first evaluate *c*1, *c*2, *c*3, and *c*4 individually. Then we apply *and*s and evaluate (c1 and c2) and (c3 and c4); eventually, we apply the *or* to evaluate the whole condition.

Because our parsing algorithm is *recursive*, it starts with the lowest precedence (*or*), and from there it moves on to the precedence immediately above (*and*), and from there to highest precedence (elementaries). Ignoring parentheses for now, it could be as follows. This implementation is not final and will be revisited in a moment when we deal with parentheses.

```
static ExprTree parseCondL2( TokIt& cur, const TokIt end)
{
  // First parse the leftmost elem condition
  auto lhs = parseCondElem( cur, end);

  // Do we have an 'and'?
  while( cur != end && *cur == "AND")
  {
    // Advance cur over 'and' and parse the rhs
    ++cur;

    // Should not stop straight after 'and'
    if( cur == end)
    throw script_error( "Unexpected end of statement");

    // Parse the rhs elem condition
    auto rhs = parseCondElem( cur, end);

    // Build node and assign lhs and rhs as its arguments,
    // store in lhs
    lhs = buildBinary<NodeAnd>( lhs, rhs);
  }

  // No more 'and',
  // so L2 and above were exhausted,
  // return to check for an or
  return lhs;
}
```

parseCondL2() parses sequences of conditions combined with *and* until we hit an *or*, or we have consumed the condition entirely. We first parse the leftmost elementary condition into LHS with the call to *parseCondElem*(). Then, for as long as we keep hitting *and*s, we parse their RHS elementary, build an *and* node with LHS and RHS as its arguments, and store it into the new LHS. So LHS always contains the tree built so far, to be combined with another condition RHS if we hit an *and*. When we no longer hit an *and*, we just return LHS.

In something like "c1 and c2 or c3 and c4," we start the recursion with the parsing of *c*1 into LHS. Then we hit the first *and* and parse *c*2 into RHS. An *and* node is built with the LHS *c*1 and the RHS *c*2 as its arguments, and this node becomes the new LHS. For as long as we hit another *and*, this recursion is repeated and the top node of the former iteration becomes the LHS for the new iteration. When we hit something else than an *and*, like an *or* in our example, the recursion terminates and the latest result (stored in LHS at this point) is returned.

This recursion consumes an arbitrary sequence of elementary conditions combined with *and*s. When it exits, we know that we hit something that is not an *and*. This can mean the whole condition is consumed, or that we hit an *or*. This is why this function must be called from a lower level *parseCondL*1() that deals with *or*s. We simply call it *parseCond*(), since this is the lowest level for conditions.

parseCond() is identical to *ParseCondL2()*, except that it calls *Parse CondL2()* instead of *ParseCondElem()* to build its LHS and RHS, and checks for *or* instead of *and*. In fact, *ParseCond()* is so similar to *parseCondL2()* that the code may have been templated. We refrain from abusing templates here only for pedagogical reasons.

```
static ExprTree parseCond( TokIt& cur, const TokIt end)
{
  // First exhaust all L2 (and) and above (elem)
  // conditions on the lhs
  auto lhs = parseCondL2( cur, end);

  // Do we have an 'or'?
  while( cur != end && *cur == "OR")
  {
    // Advance cur over 'or' and parse the rhs
    ++cur;

    // Should not stop straight after 'or'
    if( cur == end)
      throw script_error( "Unexpected end of statement");

    // Exhaust all L2 (and) and above (elem)
    // conditions on the rhs
    auto rhs = parseCondL2( cur, end);

    // Build node and assign lhs and rhs as its arguments,
    // store in lhs
    lhs = buildBinary<NodeOr>( lhs, rhs);
  }

  // No more 'or', and L2 and above were exhausted,
  // hence condition is complete
  return lhs;
}
```

The call to *parseCondL*2() on the first line exhausts all the *and* and elementaries on the left and stores the corresponding tree in LHS. Then, for as long as we keep hitting *or*, we exhaust the *and* and elementaries on its right, and build the *or* node that becomes the new LHS, and loop.

In our example “c1 and c2 or c3 and c4,” the first call to *parseCondL*2() parses “c1 and c2” into LHS. Then we hit an *or* and perform a second call to *parseCondL*2() in order to parse “c3 and c4” into RHS. An *or* node is built with LHS "c1 and c2" and RHS "c3 and c4" as its arguments and stored into LHS in case we would hit another *or*. This is not the case in our example, and we already know we are not over an *and* either, otherwise *parseCondL2()* would not have exited. Hence, we exhausted all condition operators, which means we are done with this condition entirely. If we had hit another *or*, we would have looped until all *or*s, *and*s, and elementaries are consumed.

This multi-level recursive structure is the core of the recursive descent algorithm. Readers unfamiliar with recursive patterns may need some time to wrap their minds around this design. We strongly recommend taking this time. We use recursive programming everywhere with scripting languages, in the parsing of conditions, the upcoming parsing of expressions, and in the implementation of the visitors that perform actions based on trees before and during the Monte-Carlo simulations.

This code works without parentheses. Parentheses, however, supersede conventional precedence and must be addressed accordingly. Readers should try to find the solution themselves before reading on.

First, we need a helper function to identify the matching closing ')' when we hit an opening '(,' while skipping through nested '()' pairs. We develop a generic function that may also be used with the likes of [], , '' or "", should we require support for these in the future. And, since we know at compile time what characters we look for, we pass them as non-type template arguments.

```
// Find matching closing char, for example matching ) for a (,
// skipping through nested pairs
// does not change the cur iterator,
// assumed to be on the opening,
// and returns an iterator on the closing match
template <char OpChar, char ClChar>
static TokIt findMatch( TokIt cur, const TokIt end)
{
  unsigned opens = 1;

  ++cur;
  while( cur != end && opens > 0)
  {
    opens +=
       ((*cur)[0] == OpChar) - ((*cur)[0] == ClChar);
    ++cur;
  }

  if( cur == end && opens > 0)
  throw
     script_error( (string( "Opening ") + OpChar +
        " has not matching closing " + ClChar).c_str());

  return --cur;
}
```

This code is self-explanatory. We had an L1 parser for *ors*, L2 for *ands*, and L3 for elementaries. Parentheses introduce another "L2.5" level just below elementaries, since operations between parentheses take precedence.

Hence, *parseCond*() is unchanged but the calls to *parseCondElem*() in *parseCond2*() must be changed to a call to *parseCondParentheses*():

```
static ExprTree parseCondL2( TokIt& cur, const TokIt end)
{
 // First parse the leftmost elem
 // or parenthesed condition
 auto lhs = parseCondParentheses( cur, end);

 // Do we have an 'and'?
 while( cur != end && *cur == "AND")
 {
  // Advance cur over 'and' and parse the rhs
  ++cur;

  // Should not stop straight after 'and'
  if( cur == end)
     throw script_error( "Unexpected end of statement");

  // Parse the rhs elem or parenthesed condition
  auto rhs = parseCondParentheses ( cur, end);

  // Build node and assign lhs and rhs as its arguments,
  // store in lhs
  lhs = buildBinary<NodeAnd>( lhs, rhs);
 }

 // No more 'and',
 // so L2 and above were exhausted,
 // return to check for an or
 return lhs;
}
```

This is the only change in the existing code, but now we need to develop *parseCondParentheses*():

```
static ExprTree parseCondParentheses( TokIt& cur, const TokIt end)
{
 ExprTree tree;

 // Do we have an opening '('?
 if( *cur == "(")
 {
  // Find match
  TokIt closeIt = findMatch<'(',')'>( cur, end);

  // Parse the parenthesed condition,
  // including nested parentheses,
  //  by recursively calling the parent parseCond
  tree = parseCond( ++cur, closeIt);
```

```
        // Advance cur after matching )
        cur = ++closeIt;
    }
    else
    {
        // No (, so leftmost we move one level up
        tree = parseCondElem( cur, end);
    }

    return tree;
}
```

We replaced the call to *parseCondElem*() in *parseCondL2*() by a call to *parseCondParentheses*(). If the leftmost token is not an opening '(,' then *parseCondParentheses*() just calls through *parseCondElem*() and nothing changes. If however we do hit a '(,' then we parse the condition between parentheses with a recursive call to the parent parser *parseCond*(), but passing the matching closing ')' as the end token so that only the condition between the parentheses is parsed. Note that the recursion will take proper care of nested parentheses, too. The function will return the top node of the whole tree parsed from the entire condition between parentheses, including the nested ones.

And just like that, we added support for parentheses, and our condition parser is complete.

A.4 RECURSIVELY PARSING EXPRESSIONS

What remains is the expression parser *parseExpr*(). Good news is, it is virtually identical to the condition parser. We have more operators, hence more levels, but the algorithm and the structure of the code are the same. Readers who understood the parsing of conditions will have no difficulty with the expression code. Parsing levels for expressions are listed next, from lowest to highest precedence.

1. Addition and subtraction
2. Multiplication and division
3. Power. We use the notation "a ^ b" for "a to the power b"
4. Unary + and –
5. Parentheses
6. Functions (both mathematical and financial), variables, and constants

As for conditions, the entry point is L1 and the algorithm works its way up to L6, but we develop the code in the reverse order, starting with functions, variables, and constants, which play the same role in the expression parser as the elementary conditions in the condition parser.

```
static ExprTree parseVarConstFunc( TokIt& cur, const TokIt end)
{
 // First check for constants,
 // if the char is a digit or a dot,
 // then we have a number
 if( (*cur)[0] == '.'
    || ((*cur)[0] >= '0' && (*cur)[0] <= '9'))
 {
  return parseConst( cur);
 }

 // Check for functions,
 // including those for accessing simulated data
 ExprTree top;
 unsigned minArg, maxArg;
 if( *cur == "SPOT")
 {
  top = make_base_node<NodeSpot>();
  minArg = maxArg = 0;
 }
 else if( *cur == "LOG")
 {
  top = make_base_node<NodeLog>();
  minArg = maxArg = 1;
 }
 else if( *cur == "SQRT")
 {
  top = make_base_node<NodeSqrt>();
  minArg = maxArg = 1;
 }
 else if( *cur == "MIN")
 {
  top = make_base_node<NodeMin>();
  minArg = 2;
  maxArg = 100;
 }
 else if( *cur == "MAX")
 {
  top = make_base_node<NodeMax>();
  minArg = 2;
  maxArg = 100;
 }
 // ...

 if( top)
 {
  string func = *cur;
  ++cur;

  // Matched a function, parse its arguments and check
  top->arguments = parseFuncArg( cur, end);
  if( top->arguments.size() < minArg
     || top->arguments.size() > maxArg)
       throw script_error(
          (string( "Function ") + func
             + ": wrong number of arguments")
                .c_str());
```

```
    // Return
    return top;
  }

  // When everything else fails,
  // we have a variable
  return parseVar( cur);
}
```

First, we check if we have a constant number. This is the case if the first character is a digit or a dot. In this case we call the constant parser *parseConst*() and return.

parseConst() is self-explanatory:

```
static ExprTree parseConst( TokIt& cur)
{
  // Convert to double
  double v;
  try
  {
    v = stod( *cur);
  }
  catch( const exception&)
  {
    throw script_error
        ( (*cur + " is not a number").c_str());
  }

  // Build the const node
  auto top = make_node<NodeConst>();
  top->val = v;

  // Advance over var and return
  ++cur;
  return move( top);
  // Explicit move is necessary
  // because we return a base class pointer
}
```

If we don't hit a number, then we check against the names of all the functions in our syntax, including the ones for accessing simulated data (only *spot* for now). We only develop *log*, *sqrt*, *min*, and *max* below, although we encourage readers to implement more mathematical and financial functions at this stage. All it takes is to declare the corresponding concrete node type and mimic the code for our four functions in the parser.

If we have a match on a function, we build a node of the corresponding type, and record the minimum and maximum number of permitted arguments for syntax checking. Then we call a dedicated argument parser *parseFuncArg*(), which returns the array of arguments parsed into expression trees directly into the arguments of the newly created node. Finally, we check that the number of arguments is correct and return the node.

If we don't have a match for a function, all pattern matching has failed, meaning, we have a variable. In this case, we call our variable parser *parseVar*(), implemented earlier.

Our L6 parser is complete once we have the argument parser. It parses all arguments between parentheses following the function, separated by commas, into an array of trees.

```
static vector<ExprTree> parseFuncArg( TokIt& cur, const TokIt end)
{
  // Check that we have a '(' and something after that
  if( (*cur)[0] != '(')
    throw script_error( "No opening ( following function name");

  // Find matching ')'
  TokIt closeIt = findMatch<'(',')'>( cur, end);

  // Parse expressions between parentheses
  vector<ExprTree> args;
  ++cur; // Over '('
  while( cur != closeIt)
  {
    args.push_back( parseExpr( cur, end));
    if( (*cur)[0] == ',') ++cur;
    else if( cur != closeIt)
      throw script_error( "Arguments must be separated by commas");
  }

  // Advance and return
  cur = ++closeIt;
  return args;
}
```

We check that we start with a '(,' find the matching ')' with our helper *findMatch*() that skips over nested '),' and advance the token iterator over '(.' Each argument consists of exactly one expression, so we iterate recursive calls to the parent *parseExpr*() to parse them until we hit the matching ').' In between calls to *parseExpr*(), we check that we have a comma, and skip over it. Once we have parsed all expressions separated by commas between the parentheses that follow the function name, we return them in a vector. Note that C+11 moves the returned vector so no copies are made.[1]

Just below constants, variables, and functions, we have parentheses. The code for parenthesed expressions is identical to parenthesed conditions, except the two functions to call are *parseExpr*() (the lowest level expression parser) instead of *parseCond*() (the lowest level condition parser) if

[1]Efficient compilers are most likely to perform the Return Value Optimization, or RVO, meaning that the vector will be created on the stack of the caller, instead of being created inside the function and copied or moved to the caller.

we match '(;' and *parseVarConstFunc*() (the next level expression parser) in place of *parseCondElem()* (the next level condition parser) if we don't match '(.' Hence, we called the same function, renamed simply *parseParentheses*(), and parameterized it with the two functions to call on match/not match. The two functions to call are known at compile time, so we made them non-type template parameters:

```
typedef ExprTree (*ParseFunc) ( TokIt&, const TokIt);

template <ParseFunc FuncOnMatch, ParseFunc FuncOnNoMatch>
static ExprTree parseParentheses( TokIt& cur, const TokIt end)
{
  ExprTree tree;

  // Do we have an opening '('?
  if( *cur == "(")
  {
    // Find match
    TokIt closeIt = findMatch<'(',')'>( cur, end);

    // Parse the parenthesed condition/expression,
    // including nested parentheses,
    // by recursively calling the parent parseCond/parseExpr
    tree = FuncOnMatch( ++cur, closeIt);

    // Advance cur after matching )
    cur = ++closeIt;
  }
  else
  {
    // No (, so leftmost we move one level up
    tree = FuncOnNoMatch( cur, end);
  }

  return tree;
}
```

Note we need to modify ParseCondL2 again:

```
static ExprTree parseCondL2( TokIt& cur, const TokIt end)
{
  // First parse the leftmost elem or parenthesed condition
  auto lhs = parseParentheses<parseCond,parseCondElem>( cur, end);

  // Do we have an 'and'?
  while( cur != end && *cur == "AND")
  {
    // Advance cur over 'and' and parse the rhs
    ++cur;

    // Should not stop straight after 'and'
    if( cur == end)
       throw script_error( "Unexpected end of statement");

    // Parse the rhs elem or parenthesed condition
```

```
    auto rhs = parseParentheses<parseCond,parseCondElem>( cur, end);

    // Build node and assign lhs and rhs as its arguments,
    // store in lhs
    lhs = buildBinary<NodeAnd>( lhs, rhs);
  }

  // No more 'and',
  // so L2 and above were exhausted,
  // return to check for an or
  return lhs;
}
```

Next level 4 is for unary operators '+' and '–.' We want support for expressions like "x * -y" and even for sequences of unary operators such as in "x * –y."

```
static ExprTree parseExprL4( TokIt& cur, const TokIt end)
{
  // Here we check for a match first
  if( cur != end && ((*cur)[0] == '+' || (*cur)[0] == '-'))
  {
    // Record operator and advance
    char op = (*cur)[0];
    ++cur;

    // Should not stop straight after operator
    if( cur == end) throw script_error
       ( "Unexpected end of statement");

    // Parse rhs, call recursively
    // to support multiple unaries in a row
    auto rhs = parseExprL4( cur, end);

    // Build node and assign rhs as its (only) argument
    auto top = op == '+'?
       make_base_node<NodeUplus>():
       make_base_node<NodeUminus>();
    top->arguments.resize( 1);
    // Take ownership of rhs
    top->arguments[0] = move( rhs);

    // Return the top node
    return top;
  }

  // No more match,
  // we pass on to the L5 (parentheses) parser
  return parseParentheses<parseExpr,parseVarConstFunc>
     ( cur, end);
}
```

Like the parentheses parser, the unary operator parser starts with a check for the operators '+' or '–' on the leftmost token of the expression, and, if none is found, it passes execution through to the next level, that is, parentheses.

If a match is found, then *parseExprL4*() calls itself recursively on the RHS of the operator so that we support sequences of unary operators in a row. Then it builds the top node of the unary type corresponding to the matched operator, records RHS at its argument, and returns it. The recursion guarantees that a sequence of unary operators '+' and '−' is fully parsed on exit and the corresponding tree correctly built and its top node returned.

Eventually, algorithms for binary operators L1–L3 are identical to one another and identical to the L1 and L2 condition parsers. The level n parser, like the L1 and L2 condition parsers, steps through the following sequence:

1. Parse the leftmost expression with the level $n+1$ parser and record the tree into LHS.
2. For as long as we keep hitting level n operators:
   (a) Parse the RHS of the operator with the level $n+1$ parser.
   (b) Build a node of the type corresponding to the matched operator.
   (c) Record current LHS and RHS trees as the node's arguments.
   (d) Make this node the new LHS.
3. Return LHS.

```
// Parent, Level1, '+' and '-'
static ExprTree parseExpr( TokIt& cur, const TokIt end)
{
  // First exhaust all L2 ('*' and '/')
  // and above expressions on the lhs
  auto lhs = parseExprL2( cur, end);

  // Do we have a match?
  while( cur != end
      && ((*cur)[0] == '+' || (*cur)[0] == '-'))
  {
    // Record operator and advance
    char op = (*cur)[0];
    ++cur;

    // Should not stop straight after operator
    if( cur == end)
        throw script_error( "Unexpected end of statement");

    // Exhaust all L2 ('*' and '/')
    // and above expressions on the rhs
    auto rhs = parseExprL2( cur, end);

    // Build node and assign lhs and rhs as its arguments,
    // store in lhs
```

```
    lhs = op == '+'?
      buildBinary<NodeAdd>( lhs, rhs) :
      buildBinary<NodeSubtract>( lhs, rhs);
  }

  // No more match, return lhs
  return lhs;
}

// Level2, '*' and '/'
static ExprTree parseExprL2( TokIt& cur, const TokIt end)
{
  // First exhaust all L3 ('^')
  // and above expressions on the lhs
  auto lhs = parseExprL3( cur, end);

  // Do we have a match?
  while( cur != end
    && ((*cur)[0] == '*' || (*cur)[0] == '/'))
  {
    // Record operator and advance
    char op = (*cur)[0];
    ++cur;

    // Should not stop straight after operator
    if( cur == end)
      throw script_error( "Unexpected end of statement");

    // Exhaust all L3 ('^') and above expressions on the rhs
    auto rhs = parseExprL3( cur, end);

    // Build node and assign lhs and rhs as its arguments,
    // store in lhs
    lhs = op == '*'?
      buildBinary<NodeMult>( lhs, rhs) :
      buildBinary<NodeDiv>( lhs, rhs);
  }

  // No more match, return lhs
  return lhs;
}

// Level3, '^'
static ExprTree parseExprL3( TokIt& cur, const TokIt end)
{
  // First exhaust all L4 (unaries)
  // and above expressions on the lhs
  auto lhs = parseExprL4( cur, end);

  // Do we have a match?
  while( cur != end && (*cur)[0] == '^')
  {
    // Advance
    ++cur;
```

```
        // Should not stop straight after operator
        if( cur == end)
            throw script_error( "Unexpected end of statement");

        // Exhaust all L4 (unaries)
        // and above expressions on the rhs
        auto rhs = parseExprL4( cur, end);

        // Build node and assign lhs and rhs as its arguments,
        // store in lhs
        lhs = buildBinary<NodePow>( lhs, rhs);
    }

    // No more match, return lhs
    return lhs;
}
```

This recursive logic is identical to conditions,[2] and it has been extensively explained in our parsing of conditions. On each level, the recursion consumes all higher level expressions on the left and right of the operator, until no higher level operator is found, at which point it builds a new node depending on the matched operator, sets its arguments to the higher level LHS and RHS, makes it the new LHS, and loops for as long as we keep hitting operators for this level. When it is no longer the case, the LHS from the last iteration is returned. At this point, exactly one whole expression has been fully parsed into a tree.

A.5 PERFORMANCE

We never found parsing to be a bottleneck, even with large sets of long scripts for complex transactions. This is especially the case because parsing is only performed once, hence its cost is typically negligible with respect to the generation and the evaluation of scenarios. It is in the evaluation that performance must be optimized to the limit, since it is performed repeatedly in a potentially large number of scenarios. Parsing and visits outside of simulations do not require the same attention.

If however you need for some reason to further improve the performance of the parser, note that a very substantial speedup can be achieved without much difficulty with multi-threading and memory pools.

[2]Actually, one difficulty specific to operators '–' and '/' requires extra care. These operators used in sequence are conventionally subject to an extra precedence rule, the LHS precedence. For example, "a-b-c" must be read "(a-b)-c" not "a-(b-c)." Readers can check that our code respects LHS precedence.

Our entry point into parsing in our *Product* class iterates over event strings and parses them sequentially.

```
// Build events out of event strings
template<class EvtIt>
// Takes begin and end iterators on pairs of dates
// and corresponding event strings
//  as from a map<Date,string>
void parseEvents( EvtIt begin, EvtIt end)
{
  // Copy event dates and parses event strings sequentially
  for( EvtIt evtIt = begin; evtIt != end; ++evtIt)
  {
    // Copy event date
    myEventDates.push_back( evtIt->first);
    // Parse event string
    myEvents.push_back( parse( evtIt->second));
  }
}
```

This is what parallel programmers call an "embarrassingly parallel" problem, which means a problem that can be trivially parallelized over events. This is made even simpler with the addition of a standard threading library in C++11, so you can dispatch the parsing of event strings across the multiple processors and cores in your computer, without the need to link to any third-party library.

We refer to the excellent C++ Concurrency in Action textbook by Anthony Williams [28] for multi-threaded programming in C++11.

In addition, our algorithm keeps allocating small bits of memory on every elementary operation for the node and for its vector of arguments. The default allocators behind the new operator and the STL's vector are not optimally suited for the allocation of a large number of small bits of memory. For maximum performance, you may want to use a memory pool instead. Memory pools pre-allocate large chunks of memory, fill them with small objects as they come, and free all memory at once when it is no longer needed. Memory pools can even speed up the Monte-Carlo simulations: with memory pools, nodes are not scattered throughout memory; they remain localized together, improving the efficiency of the CPU cache.

Bibliography

[1] J. Andreasen. Cva on an ipad mini, part 3: Xva algorithms. Aarhus Kwant Factory PhD Course, 2014.

[2] J. Andreasen. Malliavin greeks for dummies. Danske Bank internal memo.

[3] L. Bergomi. *Stochastic Volatility Modeling*. Chapman and Hall, 2016.

[4] L. Bergomi. *Theta (and other greeks): Smooth barriers*. QuantMinds, Lisbon, 2018.

[5] F. Black and M. Scholes. The pricing of options and corporate liabilities. *Journal of Political Economy*, 81(3):637–654, 1973.

[6] J. F. Carriere. Valuation of the early-exercise price for options using simulations and nonparametric regression. *Insurance: Mathematics and Economics*, 19(1):19–30, 1996.

[7] J. C. Cox, S. A. Ross, and M. Rubinstein. Option pricing: A simplified approach. *Journal of Financial Economics*, 7(3):229–263, 1979.

[8] S. Dalton. *Financial Applications Using Excel Add-in Development in C / C++*. Wiley Finance, 2007.

[9] B. Dupire. Pricing with a smile. *Risk*, 7(1):18–20, 1994.

[10] B. Dupire. A unified theory of volatility. Working Paper, 1996.

[11] H.-J. Flyger, B. Huge, and A. Savine. Practical implementation of AAD for derivatives risk management, XVA and RWA. Global Derivatives, 2015.

[12] E. Gamma, R. Helm, R. Johnson, and J. Vlissides. *Design Patterns: Elements of Reusable Object-oriented Software*. Addison-Wesley Longman Publishing, 1995.

[13] J. Gatheral. *The Volatility Surface: A Practitioner's Guide*. Wiley Finance, 2006.

[14] M. Giles and P. Glasserman. Smoking adjoints: Fast evaluation of greeks in Monte Carlo calculations. *Risk*, 2006.

[15] P. Glasserman. *Monte Carlo Methods in Financial Engineering*. Springer, 2003.

[16] J. Guyon and P. Henry-Labordere. The smile calibration problem solved. *SSRN*, 2011. Available at SSRN: https://ssrn.com/abstract=1885032 or http://dx.doi.org/10.2139/ssrn.1885032.

[17] P. Henry-Labordere. Counterparty risk valuation: A marked branching diffusion approach. *The Review of Financial Studies*, 2012. Available at SSRN: https://ssrn.com/abstract=1995503 or http://dx.doi.org/10.2139/ssrn.1995503.

[18] B. N. Huge and A. Savine. Lsm reloaded—differentiate xva on your ipad mini. *SSRN*, 2017.

[19] P. Jackel. *Monte Carlo Methods in Finance*. Wiley Finance, 2002.

[20] B. Leif, G. Andersen, and Vladimir V. Piterbarg. *Interest Rate Modeling, Volume 1, 2, and 3*. Atlantic Financial Press, 2010.

[21] A. Lipton. *Mathematical Methods for Foreign Exchange*. World Scientific, 2001.

[22] F. A. Longstaff and E. S. Schwartz. Valuing American options by simulation: A simple least-square approach. *The Review of Financial Studies*, 14(1):113–147, 2001.

[23] H. McKean. Application of Brownian motion to the equation of Kolmogorov-Petrovskii-Piskunov. *Communications on Pure and Applied Mathematics*, 28(3):323–331, 1975.

[24] U. Naumann. *The Art of Differentiating Computer Programs: An Introduction to Algorithmic Differentiation*. SIAM, 2012.

[25] W. H. Press, S. A. Teukolsky, W. T. Vetterling, and B. P. Flannery. *Numerical Recipes: The Art of Scientific Computing*. Cambridge University Press, 2007.

[26] A. Savine. Volatility. Lecture Notes from University of Copenhagen, tinyUrl.com/ant1savinePub password: 54v1n3 folder: Vol.

[27] A. Savine. *Modern Computational Finance: AAD and Parallel Simulations*. Wiley Finance, 2018.

[28] A. Williams. *C++ Concurrency in Action*. Manning, 2012.

[29] L. A. Zadeh. *Fuzzy Sets, Fuzzy Logic, and Fuzzy Systems, Selected Papers by Lotfi A. Zadeh*. World Scientific, 1949–1995.

Index